WHY ELEPHANTS HAVE BIG EARS

CHRIS LAVERS

ST. MARTIN'S PRESS NEW YORK

WHY ELEPHANTS HAVE BIG EARS

UNDERSTANDING PATTERNS OF LIFE ON EARTH

 For information, address St. Martin's Press, 175 Fifth Avenue, New York, N.Y. 10010.

www.stmartins.com

Library of Congress Cataloging-in-Publication Data

Lavers, Chris.
Why elephants have big ears : understanding patterns of life on Earth / [Chris Lavers].
p. cm.
Includes bibliographical references (p.).
ISBN 0-312-26902-1
1. Animal ecology. 2. Animals—Adaptation. 3. Evolution (Biology) I. Title.
QH541.L352 2001
590—dc21 00-045997

First published in Great Britain by Victor Gollancz, an imprint of Orion Books Ltd.

First U.S. Edition: March 2001

10 9 8 7 6 5 4 3 2 1

FOR SOMAI

CONTENTS

ACKNOWLEDGMENTS

I wanted early drafts of this book to be reviewed by scientists, writers, and enthusiasts who did not know me personally. The subject matter ranges across several disciplines, some of them no longer my own, so I felt that the manuscript needed serious criticism. I rattled off E-mails to some eminent authorities requesting their help, and it says a lot about the nature of my profession that I did not receive a single refusal. Some offered to review one or a few chapters in their particular areas of expertise, while others generously undertook the onerous task of reviewing the whole book. No-holds-barred criticism of the most constructive kind was exactly what I got back, and the book is immeasurably better for it. My gratitude to Sallie-Anne Bailey, David Bellamy, Jason Bourke, Chris Brochu, Martin Human, Stephen Hurrell, Hilary, Melanie, and Michael McCullagh, Richard Morgan, Ilja Nieuwland, David Norman, Sarah O'Hara, Gregory S. Paul, the greatest living animal physiologist, who, as is his wont, refuses to be acknowledged for his kindness, Dan Tapster, Ruth Tingay, and Patrick Walsh. Any remaining errors of fact, reasoning, or inference are mine.

Any wildlife artist working on a standard pay-per-picture basis will know that mammals and birds aren't worth the money. Evolution has furnished these creatures with very fiddly insulating coverings which take forever to fill in. Heartfelt thanks to Somai Man for helping with the furs and feathers. Thanks also to Chris Lewis for his skill in drafting the maps and graphs.

Thanks to Sara Holloway (Gollancz), Joshua Kendall (St. Martin's), Ingrid von Essen, and Brenda Woodward for their editing skills, and to Patrick Walsh for advice, encouragement, and business acumen.

Finally, I owe a debt of gratitude to the students at the University of Nottingham—particularly those in the School of Geography—who, with patience and good humor over many years, have put up with me trying out ideas on them. This book consists largely of those that they didn't laugh at.

PREFACE

This book was born of ignorance and confusion. It was conceived while puzzling over wildlife documentaries as a child, most of them made by the British Broadcasting Corporation's Natural History Unit, a group of internationally renowned researchers, cameramen, and camerawomen fronted by that inspirational whispering narrator, David Attenborough. Like millions of British children I was enthralled and awestruck at the beauty of the natural world revealed by Attenborough's documentaries, but they were also the cause of much frustration. His camera crews seemed to follow big game and their predators endlessly around the African Serengeti, recording in loving detail every kill, sexual conquest, play-fight, and bout of communal grooming, but, to my memory at least, that was about as far as it ever went. Granted, David would expound the odd scientific theory from time to time, but he never seemed to grapple with the really *obvious* questions.

Why, for instance, were all the large animals caught on camera on the savannas of Africa mammals? Elephants, rhinos, giraffes, zebras, cats, hyenas, hunting dogs, and a bewildering variety of antelopes were all captured in glorious technicolor by Attenborough and his colleagues for our armchair edification, but no big reptiles or amphibians were anywhere to be seen. This was particularly puzzling because I knew that dinosaurs were reptiles, and some of these magnificent beasts had been famously huge. Why did big terrestrial reptiles exist in the Cretaceous period, yet not now in Africa? Moreover, it was clear from my animal encyclopedia that big land-living reptiles were rare nearly everywhere else too. The complement for the entire planet seemed to consist of five snakes, some giant tortoises, and a buffalo-eating lizard from the Malay archipelago. Why so few? And just to add to the confusion, it was equally clear from David's documentaries that the most characteristic big animals in rivers and lakes were crocodilians and turtles of various sorts. Big mammals dominate the land, while big reptiles dominate the water. Why?

Perhaps, I thought, it had something to do with the fact that mammals are warm-blooded and reptiles cold-blooded. Maybe the heat-sapping effect of water makes life difficult for animals with hot bodies. Could this be why African rivers are full of crocodiles and not antelopes? At first this seemed like a good idea, but then my encyclopedia clearly showed that the world's oceans were full of warm-blooded seals and whales. So freshwater favors cold-bloods, while saltwater favors warm-bloods. This didn't make sense at all.

And to make matters even more complicated, these patterns seemed to apply only to animals above a certain size. When David and Co. stuck their macro lenses under rocks, in cracks, down burrows, or into tree holes, all sorts of four-legged animals came into view. The nooks and crannies of tropical forests turned out to house a plethora of small cold-blooded frogs, lizards, and snakes alongside all the warm-blooded songbirds, rodents, and shrews. Cool, wet places seemed to be overrun with little amphibians. Deserts contained all manner of tiny reptiles. Terrestrial cold-bloods become more successful in the struggle for life as they get smaller. Why?

I went to the library fully expecting to find straightforward explanations of why mammals are best at being big on land, why reptiles lord it as big animals in freshwater, why these patterns of dominance change with scale, and so on. But the natural-history books in my local library didn't say. I began to think I was just dim and the answers so obvious that no one bothered mentioning them anymore, not even David. Perhaps they had been worked out so long ago that you had to dig up the really old literature to find out. I made a mental note to nose around a college library in the unlikely event that I ever managed to get inside one.

And eventually I did. But the books in my college library weren't much help either. Mostly they would plunge straight into the minutiae of some specialist subject—animal physiology, biophysics, genetics, population ecology, paleoecology—and leapfrog the really obvious questions in the process. There were some interesting clues in some of the books, but no one was tackling the issues head-on. Had biologists really rushed headlong into their various areas of specialization in the first half of the twentieth century without pausing to finish off the Victorian discipline of natural history first? I must have spent hundreds of hours looking for answers. Little by little the clues accumulated, and some ideas about possible solutions to those nagging childhood questions began to emerge. And, as is always the way in science, the deeper I looked, the more puzzles came to light, so I ended up having to consult books and wiser heads to find the answers to many grown-up questions too. The result of this twenty-five-year period of intellectual groping is the slim volume that you are currently reading.

Largely because of my own struggles with the primary scientific literature while researching and writing this book, I have attempted to write as simply as possible with the minimum of mathematics and jargon (and for American readers of this St. Martin's Press version, I have used imperial measurements in strategic places as well as metric). Where I have given in and used a technical term it is simply because I cannot think of a suitable alternative. Similarly, it is customary in academic circles to reference the work of others as one goes along, a practice that often results in more space being given over to lists of names and dates than to ideas and arguments. This approach is so ingrained in the scientific community and becomes so natural after a while that abandoning it makes one feel rather uneasy, but for the sake of clarity and brevity I have done so. Anyone wishing to chase up the primary literature should refer to the section entitled "Key References and Selected Further Reading," which contains a chapter-by-chapter, subject-by-subject guide to useful background material. My fondest hope is that children of school age will read this book and want to learn more, so I have coded each reference in the Bibliography from 1 to 3 according to degree of difficulty: 1 is for anyone, 2 is for the serious, and 3 is for the brave.

The most satisfying aspect of researching the topics covered herein has been the constant confirmation that many of the broad patterns of life on our planet can be explained using just a few simple biological principles. Chief among these is that an animal's fuel requirements crucially affect its lifestyle. Fuel consumption varies enormously between different types of animals as well as between similar types of animals of different sizes, and although creatures go about the business of garnering and rationing their energy in a bewildering variety of ways, the Darwinian bottom line is nearly always the same: be good at it or make way for something better. But the best way of fueling a life also varies from place to place and from environment to environment, and some types of animals, because of the twists and turns of their complicated evolutionary history, have acquired the bodily equipment to triumph under some circumstances but not others. And so there is a natural order to life, underpinned by the pros and cons of different methods of manipulating energy, with one type of creature over here doing such and such, and another over there doing something rather different. Darwin's process of evolution by natural selection has given rise both to the staggering diversity of life on our planet *and* the beautiful order in which it is arrayed, and this book sets out to explore, and unashamedly revel in, both.

Back to the sea again! Down there today I watched the goings-on of the sea-snails, patellas (mussels with a single *shell), and the pocket-crabs, and it gave me a glow of pleasure observing them. No, really, how delightful and magnificent a* living thing *is! How exactly matched to its condition, how true, how intensely* being*! And how much I'm helped by the small amount of study I've done and how I look forward to taking it further!*

—JOHANN WOLFGANG VON GOETHE,
diary entry, October 9, 1786

WHY ELEPHANTS HAVE BIG EARS

[1]

WHY ELEPHANTS HAVE BIG EARS

Consider the oddness of elephants. At between 4 and 7 metric tons (U.S. equivalents 4.4 and 7.7 tons) these creatures are fully twice as heavy as any other land animal on Earth. They have 3-meter-long noses (almost 10-feet). The African variety has the largest earflaps of any animal in history. Nearly all land mammals are covered with hair, but not elephants. Their front teeth can grow to 3 meters in length and over 200 kg (440 pounds) in weight. Consider an animal capable of scratching its knees with its teeth without bending down. We have become so familiar with elephants from zoos, books, and television documentaries that we tend to take them for granted, which is something of an achievement, all things considered.

The aim of this chapter is to explain elephants and why they evolved as they did, and not just because they are strange and fascinating animals in their own right. Elephants are the ideal starting point for an exploration of size and energy use in the animal kingdom because they are the largest creatures currently walking the planet and because their metabolic engines are among the most expensive to run. Once we appreciate the workings of these enormous gas-guzzlers, it will be but a short step to an understanding of why rats are furry, why there are no fly-sized or snake-shaped mammals, why the tiniest backboned animals are lizards and frogs, why King Kong could never have climbed the Empire State Building, and much else besides. Ultimately, a knowledge of how elephants work will lead us to the most profound disturbance to the Earth's biogeography in the last 65 million years, a man-made crisis that has left the biosphere teetering on the edge of global mass extinction. But we are jumping ahead. Our exploration of patterns of life on our planet begins with the largest land animals on Earth. And to understand these magnificent creatures we must first explore some of the biological consequences of being big.

Like most animals, elephants are an odd shape and thus rather difficult to measure; so, for the purposes of illustration, let us imagine that they are melons. A cantaloupe melon is around 16 cm (6.2 inches) in diameter or about twice the width of an orange. A linear measure such as width is one way of expressing the relative size of these two objects, but it is not the only comparison we could make. The melon, for example, has a surface area four times greater than that of the orange. Cut the fruits, and the cross-sectional area of the melon will also be four times greater. The areas of spherical objects—surface or sectional—always scale in this way: double the width, quadruple the area. The volume of the melon, however, is eight times that of the orange, and because oranges and melons are mainly water, and water weighs the same no matter where it comes from, volumes and weights naturally scale in the same way. Although it is only twice as wide, the melon is eight times as heavy as the orange and contains eight times as much juice (Fig. 1.1).

This general relationship between length, area, and weight always holds regardless of the objects involved, provided they are roughly the same shape. And the rules apply just as well to objects that grow from one size to another. How much does a 12-cm (4.7-inch) fish have to grow to double in weight? Only about 3 cm (1.2 inches). An ostrich egg is only 2.5 times the width of a hen's egg, but it would make the equivalent of a twenty-egg omelette. Small

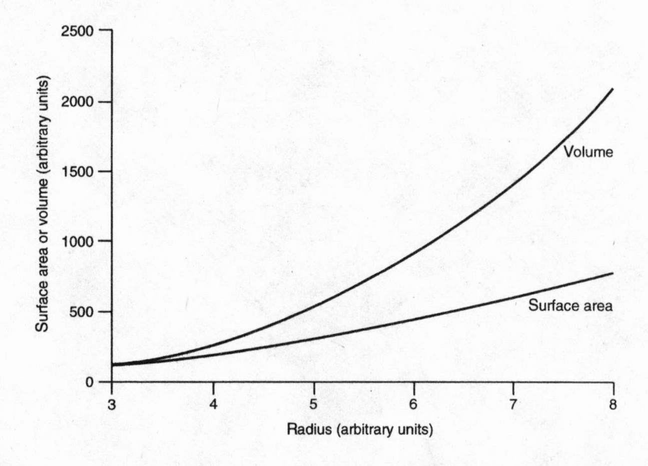

FIGURE 1.1 As the radius of a fruit, ball, or any other roughly spherical object doubles, its surface area quadruples and its volume and weight increase by a factor of eight. The factors of increase are less straightforward for irregularly shaped solids, but the general rule holds.

increases in length result in large increases in area and enormous increases in volume and weight.

How do these geometric principles affect animals? Figure 1.2 shows a close-up of an African elephant's remarkable cranial anatomy, and Figure 1.3 shows an elephant and a gazelle drawn to the same size. Many anatomic differences between the two creatures are obvious regardless of scale, but when they are standardized by height and length it is easier to make direct comparisons of the shapes and relative sizes of various body parts. Compared with the gazelle, the elephant has thicker, straighter legs, a shorter, more robust neck, a massively elongated nose, a distinct lack of hair, and, of course, those extravagant ears. Curiously, all these characteristic elephantine features are a consequence of the scaling relationship between areas and volumes.

Working from the ground up, the strength of a leg bone depends mainly on its cross-sectional area, and legs do the job of

FIGURE 1.2 The head of an African bush elephant.

supporting an animal's weight. Imagine what would happen if a gazelle remained the same shape but grew to the size of an elephant. As it doubled in height, the cross-sectional area of its bones would quadruple, but the weight of the whole animal would increase by a factor of eight. The gazelle would not have to double in height many more times before its bones would snap under the influence of gravity (although the muscles and tendons in its legs would probably give out first). Over perhaps tens of thousands of years, elephants did evolve from animals the size of gazelles, so this problem of bone strength had to be resolved. As they grew larger over evolutionary time, their shank bones thickened disproportionately fast to cope with the increasing load, which is why an elephant's legs look stockier than a gazelle's. But bulking up bones was not the whole solution: an elephant's legs are also arranged in a very unusual way, which, among other things, explains why they are nowhere near as athletic as many smaller creatures.

FIGURE 1.3 An elephant and a gazelle drawn the same size.

Gazelles have the standard arrangement of leg bones characteristic of nearly all fast-moving mammals. Their front legs look like our own: straight up and down with a joint in the middle. This middle joint, however, is not a knee but the equivalent of our wrist. The shank below this joint corresponds to the bones in the palm of our hand, while the equivalent of our elbow is right up near the animal's chest. Regardless of the bones involved, straight legs are useful because they can be locked into position with little muscular effort. But a gazelle's back legs are different. About halfway down is a backward-bending joint which is the equivalent of our ankle. The knee, again, is right up near the torso and often hidden by skin and fur. This arrangement of bones gives gazelles and many other running mammals the curious appearance of legs that bend forward at the front and backward at the rear. None of the joints in a gazelle's back legs is straight and locked like ours, which means that energy has to be expended to prevent them from collapsing.[1] If the straight-

1. Some species have a complicated arrangement of tendons, tendonlike muscles, and ligaments that allows the back legs to be semilocked when standing dead still. This mechanism has to be unhooked before the animal can move.

FIGURE 1.4 A cheetah's enormous stride-length is a product of relatively long legs and a remarkably flexible back. At full stretch the whole animal is virtually parallel to the ground.

up-and-down arrangement is more economical, why do gazelles have bent back legs?

Gazelles are magnificent runners. They have to be because super-fast predators like cheetahs regard them as little more than mobile larders (Fig. 1.4). The architecture of a gazelle's back legs is best understood by appreciating that these creatures spend the most critical moments of their lives running away. Flat-out speed is important, but for the relatively short races between gazelles and cheetahs, acceleration and cornering ability are probably more critical. Human sprinters know that they can achieve the greatest rate of acceleration by assuming a crouched position and straightening their legs explosively when the gun fires. Tenths or even hundredths of a second are crucial for sprinters, but much less so for distance athletes, which is why sprinters get down on all fours to start a race and marathon runners don't bother. Needless to say, acceleration is even more crucial if the prize is to escape a predator's jaws. The back legs of a gazelle are arranged in a permanent sprinter's crouch, allowing rapid acceleration from a standing start. The bent arrangement also means that the back legs are slightly longer than the front ones, which maximizes the length of the thrust stroke. (Other explosive accelerators like frogs and grasshoppers exploit the same principle.) In addition, when startled by a predator, a gazelle does not so much run as explode forward in a series of jumps. After each jump the back legs automatically recoil into their normal bent arrangement ready for the next one without having to be dragged all the way back into position by big heavy muscles. This system of elastic recoil saves energy, reduces leg weight, and shortens the time lag between each lifesaving thrust.

A gazelle's legs are also exceptionally thin because the leg muscles are concentrated right up near the animal's torso. Imagine starting a 100-meter sprint or negotiating a sharp corner at speed with a 5-kg (11-pound) weight tied to each ankle, and the advantages of a gazelle's top-heavy arrangement of muscles should be obvious. The legs are lightened as much as possible at the foot end, which moves through the greatest length of arc, allowing them to be quickly set in motion and maneuvered easily around corners.

So a gazelle's legs are lightly built for agility and the back ones are coiled and ready-cocked for explosive acts of locomotion. The downside of this arrangement is that muscles have to work to keep the back end of the animal from collapsing when it is just wandering around. The cost is obviously minor compared with the benefits, at least for gazelles, but the situation becomes less clear as animals get bigger. The strength of a muscle, like that of a bone, depends on its cross-sectional area, but its job is to support, lift, lower, and otherwise shift weights. If an animal were to double in height while remaining the same shape, the weight of any anatomic structure would increase roughly eightfold, while the cross-sectional area of the muscles attached to it would increase only fourfold. In other words, scale up a gazelle to the size of an elephant, and its back end would probably collapse.

FIGURE 1.5 The bent back legs of the plant-eating elephant-sized dinosaur *Triceratops* suggest that these animals may have been able to gallop rhino-style away from or toward their enemies.

There are a number of adjustments that could be made to prevent this from happening. Muscles could be pumped up to heroic proportions in order to retain the locomotory benefits of a bent back end. Some dinosaurs, like *Triceratops* (Fig. 1.5), were as heavy as elephants but had bent back legs, and some paleontologists believe that they were able to gallop in much the same way that rhinos do today. Perhaps the fact that *Triceratops* shared its world with enormous predators like *Tyrannosaurus* (Fig. 1.6) explains why powerful, thrusting back legs were a distinct advantage. The solution that elephants adopted, however, was to do away with bent back legs altogether and replace them with the straight-up-and-down variety that they have at the front. This arrangement saves a lot of muscle power but at the expense of speed and acceleration: because of their straight legs and enormous bulk, ele-

FIGURE 1.6 The main tormentor of *Triceratops* and other large late-Cretaceous vegetarians, the fearsome meat-eater *Tyrannosaurus*. Length 12 meters (39 feet).

phants are the only land mammals that cannot gallop.[2] When they run, all four legs swing backward and forward like mobile columns, which is why they look a bit more ungainly in full flight than gazelles and cheetahs. But unlike gazelles, adult elephants are not pursued by predators (except humans), and unlike cheetahs they are vegetarians, so they don't need to run toward or away from anything with any great urgency. Their particular arrangement of leg bones is probably a good compromise.

The scaling relationship between areas and volumes underlies

---

2. Because of their huge weight and inertia, elephants are also more vulnerable to certain types of misfortune than smaller creatures. As one biologist graphically put it, if an elephant were to jump like a mouse it would break its legs on takeoff and collapse them completely on impact. Stronger skeletal scaffolding allows elephants to break into a quick lumber when occasion demands, but otherwise they are rather circumspect creatures when it comes to locomotion. The old elephant enclosure at London Zoo, for example, had a shallow-sloped trench between the elephants and the public to keep the two species apart. The trench was ludicrously small, but it was quite big enough to deter the elephants from trying to cross it. Elephants in the wild negotiate quite steep slopes, but they do so very gingerly, often on their bottoms with their front legs splayed out as braces.

many of the structural and locomotory problems associated with being a large animal of any kind. As animals grow, adjustments have to be made to account for the fact that weights naturally increase faster than the strengths of muscles, tendons, and bones. For this reason if no other, we should always be suspicious of the cinematic license often taken with animals. Godzilla looks like a scaled-up *Tyrannosaurus* and King Kong like a scaled-up gorilla, but something has to give if animals are to reach such monstrous sizes. Godzilla certainly would not be able to move very fast with all that weight supported on such spindly legs, and King Kong would have had great difficulty pulling his huge body up anything with such underpowered muscles, let alone all the way up the Empire State Building. If these fantasy monsters were to work reasonably within the laws of scaling, they would have to bulk up their muscles and bones so much that they would no longer look very much like the sleek and scary animals they are supposed to represent.

The same scale-related reasoning can be used to shed light on other anatomic differences between our two chosen animals. Relative to the gazelle, for instance, the elephant's head is attached to its torso via a much shorter neck (Fig. 1.3). Now if an animal is to grow large over evolutionary time and keep its head in proportion with the rest of its body, the laws of scaling dictate that it will inevitably have to cope with a very heavy head. And the problem of head weight for elephants is exacerbated by their extraordinary teeth: the four molars at the back of an adult elephant's jaw[3] may measure 12 cm (4.7 inches) in width and 35 cm (14

3. The early ancestors of elephants had all six cheek teeth present in each quadrant of the jaw at the same time, just as most mammals do. Modern elephants still have all these teeth, but they develop sequentially throughout life. Tooth one, two, and three are used while the elephant is young. Tooth four develops at the back of the jaw and migrates forward as the earlier ones wear out and crumble away. The fifth develops at age twelve, and the final one at around twenty-five. This last set has the awesome responsibility of keeping an elephant going for the rest of its life, which may be a further fifty years. The longevity of elephants explains this odd sequential development

inches) in length and are immensely heavy. They are easily capable of pulverizing acacia branches complete with 10-cm (4-inch) thorns. The size and grinding efficiency of an elephant's teeth are crucial as these animals must consume enormous amounts of food to keep their huge bodies functioning. African bulls, for example, may consume 300 kg (660 pounds) of vegetation every day, and chewing gives an important head start to the digestive process by shredding the plant material into small bits that are more easily broken down by digestive juices in the stomach.

An elephant's back teeth may be unusual, but the front ones are quite bizarre. The upper incisors have become modified into tusks, which in the largest individuals may project 3 meters (10 feet) from the upper jaw. These impressive structures are used for sexual display, threatening rivals, fighting, or as tools working in concert with the trunk to collect and manipulate food. The largest tusks ever recorded belonged to an African bull shot near Mount Kilimanjaro in 1897 and are now on display in the Natural History Museum in London: they are each over 3 meters long and together weigh 200 kg (440 pounds).

A 1 tonne head necessarily has to be supported against the pull of gravity by very robust muscles, which is why elephants have such thick necks. To understand why their necks are also short, imagine sitting with your elbow planted firmly on a tabletop holding a heavy object like a cannonball. In this position, the full weight of the cannonball is supported mainly by the muscle that runs down the front of the upper arm and attaches to the forearm

---

of teeth. If all the teeth were present at the beginning of the animal's life, they would never stay the course. Old, worn teeth need to be replaced with new, fresh ones, and sequential development is a neat solution to the problem. It is not foolproof, however: six is the maximum number of teeth available to each quadrant of the mouth, and when the last set wears out, an elephant will inevitably starve to death. It is probably for this reason that elephants regularly clean abrasive dirt from clumps of grass before chewing them by rubbing the roots on the rough underside of their trunks.

near the elbow (the bicep). Next, imagine the cannonball fixed to a long stick and held in the same position as before. The stick effectively lengthens the forearm and puts the weight of the cannonball much further from the pivot at the elbow. Shifting the weight forward like this would increase the strain on the bicep ruinously. In general, if heavy weights have to be carried around and manipulated, it is best not to suspend them on the end of long levers. Animals with heavy heads, therefore, tend to have short necks. This is true wherever we look in the animal kingdom. Elephants, rhinos, buffaloes, and many extinct animals like mammoths and the dinosaur *Triceratops* all converged on the same principle of keeping their heavy heads on the end of short levers. Similarly, all the really huge dinosaurs that are famous for their long necks are equally famous for their relatively tiny heads.

We are now in a position to understand what is probably the strangest feature of this very unusual animal. Why do elephants have trunks? The most widely accepted answer is almost comical, and no less so because it stands a good chance of being true. Elephants probably have trunks for the simple reason that without them their heads would not reach the ground. An elephant may be 3 or 4 meters (10 to 13 feet) tall, so if it were to forage for food at or near ground level without a trunk, it would need a neck 3 to 4 meters long. It probably isn't feasible to lighten an elephant's head all that much because of the large braincase and huge jaws and teeth that it needs to process food, and we have already established why no animal in the history of life has carried a 1-tonne head on the end of a 3-meter lever. It is likely that elephants evolved from much smaller animals with prototrunks such as some tapirs have today (Fig. 1.7). As they increased in size over many thousands of years, their trunks simply lengthened to stay in contact with the ground.

Alternatively, it has long been suspected that elephants are closely related to sirenians (manatees and dugongs) and that the

FIGURE 1.7 Malayan tapir. Length around 2 meters (6.6 feet).

common ancestor of both groups may have been a fully aquatic animal of some kind. This raises the possibility that trunks developed during an earlier aquatic phase of elephant evolution and may originally have been used as snorkels.

The laws of geometry have also left other marks on the anatomy of elephants, baldness being one of the more subtle. Like you and I, elephants are mammals, and mammals are characterized by a number of traits that set us apart from the rest of our tetrapod cousins (tetrapods are four-legged animals with backbones; that is, amphibians, birds, mammals, and reptiles). We are warm-blooded, suckle our young on milk, have mouths with teeth, and bodies covered with hair. Birds are also warm-blooded, but they lay eggs, have no teeth, and are covered with feathers. Reptiles are cold-blooded with scaly skin, and amphibians are cold-blooded with smooth, moist skin. Mammalian features distinguish us from all other types of tetrapods, but this is not to say that all defining mammalian characteristics are equally obvious in every species. And hair is a good case in point. Over 99 percent of all nonaquatic mammals are furry, but elephants are all but bald. To understand why, we need to know a bit more about the nature of the mammalian metabolic engine and the way in which body size influences how it works.

Of all the tetrapods on Earth, only mammals and birds are classically warm-blooded. The members of these two great classes maintain a relatively high and constant body temperature between 30°C and 42°C (86°F and 108°F) by generating heat internally and regulating the rate at which it escapes into the environment. The heat is produced by chemical reactions that occur inside cells, and these proceed at a furious rate inside warm-blooded animals even when we are at rest or asleep. There are various ways in which mammals and birds regulate the heat produced by their tissues in order to stay at their particular set temperature. If the environment gets too hot, they can hide in cool places like burrows and divert blood to the skin where heat can escape more easily into the environment. Larger animals find it difficult to hide, but they contain a relatively large amount of water, just as melons contain a lot of juice, so they can afford to operate an evaporative cooling system—sweating or panting—to rid themselves of excess heat. (One gram of water carries away 2.4 kilojoules of heat as it evaporates. Some mammals sweat, some pant, and some are capable of both. Birds do not sweat, but most of them pant.) If the temperature of the environment drops, mammals and birds can divert blood to the center of their bodies to keep it warm and heat themselves by shivering. Shivering is not an annoying side effect of being cold, it is the body's most effective way of producing heat, because when muscles work—voluntarily or otherwise—they draw on energy supplied by elevated rates of heat-producing chemical reactions within cells. The muscular effort of shivering can raise the internal heat production of a warm-blooded animal to five times its resting level.

Cold-blooded animals, in contrast, have lower metabolic rates, produce much less heat, and cannot shiver. Warm muscles contract more efficiently and powerfully than cold ones regardless of what sort of animal they are in, so cold-bloods like lizards frequently wish to attain body temperatures in the mid-to-high 30s (95°F to 102°F) too. But because of their low metabolic rates, they have no choice but to warm themselves behaviorally, by basking in the sun, for example. Voluntary muscular exercise also produces heat, so cold-blooded animals could, in theory, heat themselves by running around all the time, but the energetic cost of such a strategy is obviously prohibitive. Pythons are a possible exception to this rule as they can elevate the temperature of their bodies in the early stages of digestion or when brooding eggs by rhythmically contracting their skeletal muscles, but this is the only known example of a cold-blooded tetrapod using muscular exercise specifically for the purpose of controlling its body temperature. (Even though snakes are legless, they are classed as tetrapods because they evolved from animals with legs. The same is true of legless mammalian tetrapods like whales and dolphins.) Cold-blooded animals also have a number of ways of cooling off if they get too hot. They can seek shade, orient their bodies to minimize the area of skin illuminated by the sun, escape to water, gape (which allows cool air to circulate around the mouth and tongue), and some can even change the color of their skins. Some lizards stand on their hind legs to catch the breeze, and a few can create their own air movements in an emergency by running around. Some arboreal species do the same by jumping or gliding between trees.

The most important difference between warm-blooded and cold-blooded animals is the primary source of heat: warm-bloods generate large amounts of heat internally, while cold-bloods rely primarily on external sources like the sun. There are some interesting cases (which we will examine later) where this dividing line becomes a little blurred, but basically mammals and birds are the only tetrapods on Earth capable of producing enough metabolic heat to raise their bodies to a high temperature without engaging in muscular exercise.

Now, maintaining a high body temperature at all times is costly. To keep a house at 38°C during a temperate winter would require the central-heating system to be turned up high and fed with a lot of fuel. So it is with warm-blooded animals. To keep their metabolic fires raging, mammals and birds consume about ten times more food than reptiles of similar size. To keep a house at 38°C but cut down on fuel bills, the only option is insulation. Investing in a vacuum-sealed boiler, installing loft insulation, and pumping foam into cavity walls are all measures that prevent heat from escaping from our houses. Warm-blooded animals work on the same principle except that they are covered with hairs or feathers that trap an insulating layer of air next to the skin. Mammals living in cold climates tend to have very dense fur to guard against heat loss. The denser the hair, the narrower each strand tends to be, and the softer the fur feels to the touch. Baby harp seals, mink, lynx, snowshoe hares, and Arctic foxes have all suffered mightily at the hands of human hunters precisely because of their luxurious insulation.

So mammals and birds produce a lot of heat, and a layer of insulation stops it from leaking out. If, say, 1000 metabolic reactions occur inside a cell every minute, then the total amount of heat produced by a large animal will be greater than that produced by a small one simply because the larger animal has more cells. But animals lose heat to the environment through their skins, and the laws of scaling dictate that volumes increase much faster than surfaces as animals get bigger. So as animals increase in size, the amount of heat-producing flesh inside them increases rapidly, but the amount of skin through which the heat can escape increases more slowly. Clearly some adjustment has to be made if large animals are to avoid cooking themselves.

As animals evolve toward large size, they cope with this problem by reducing the rate at which their cells produce heat—the larger the animal, the greater the reduction. But there is something extremely curious about this adjustment when looked at in detail. Elephant cells do produce less heat than gazelle cells, but this difference is not enough to balance an elephant's overall heat budget. And it turns out that this mismatch is characteristic of creatures of all kinds: large animals within any particular taxonomic group tend to produce more heat than expected on the basis of the scaling relationship between areas and volumes, and the larger the animal, the greater the discrepancy. Generations of biologists have banged their heads against this particular brick wall, and we still do not have an explanation. There are a number of theories, and new ones come along every few years, but none is very satisfactory. There are ways of making the mathematics work so that the mismatch can be accounted for, but there is little agreement on which way is best. Clearly, as animals evolve toward large size, they adjust their metabolic rates to keep pace not with their dwindling surfaces, but according to some other rule or rules of which we remain stubbornly ignorant.

Whatever the reason for this state of affairs, the fact is that large animals produce more heat than seems sensible. It should be clear now where this is leading. African elephants, the largest land animals on Earth, also happen to live in one of our hottest climates. Their bodies produce an enormous amount of heat, and the sun beating down on their backs doesn't help. Even though they have adjusted their thermostats to reduce the heat output of their cells, this countermeasure has fallen some way short of compensating for their relatively small surfaces. So the baldness of elephants makes sense. The last thing that such a large animal needs, especially one that lives under a baking sun most of the time, is a fur coat. White rhinos are the second-largest land animals on Earth, they also live in hot climates, and they are also bald. It is likely that such large mammals living in hot places *must* be bald in order to lose heat through their skins at an adequate rate when temperatures soar.

Hippos are also bald, but the reason in their case is less obvious. Semiaquatic animals can retreat to water if the air gets uncomfortably hot, so bald skin may not be as important for hippos as for elephants and rhinos. Some smaller artiodactyls (the taxonomic group to which hippos belong), e.g., warthogs, bison, and babirusas, are also bald or partly so, and many of the piglike members use fat for insulation rather than hair, so baldness in this group of animals is obviously not a simple function of size. The closest relative of the common hippo is the pygmy hippo, a bald animal weighing no more than 275 kg (600 pounds) and spending most of its time on land. Simple theories based on scaling arguments have to be evaluated carefully in the complex world of living things.

We are now, at last, in a position to answer the title question of this book: Why do elephants have big ears? Or, to put it in its comparative form for those who already know the answer but not the reason, why do elephants have big ears when rhinos, giraffes, and all the other warm-blooded mammals toiling under the African sun do not? The crucial difference between elephants and all other land animals is probably size: the mismatch between the amount of metabolic heat produced and the amount of skin through which it can escape tends to increase as animals get bigger, and elephants are fully twice as heavy as any of their savanna neighbors. So being bald like a rhino is not enough. Elephants need some additional way of ridding themselves of excess heat, and large flat ears densely packed with tubes through which hot blood can be pumped are a rather neat solution.

Of course, heat will not flow out of an elephant's ears if the surrounding air is warmer than 38°C unless the cooling process is aided by the evaporation of water. Unlike most other large mammals, elephants do not have the ability to sweat—possibly because their aquatic ancestors didn't need to—so they must either seek a favorable thermal gradient in the shade or find some other way of wetting their ears (and, just for good measure, other parts of their bodies too). Despite their enormous bulk, elephants regularly manage to find shady spots, and they endeavor never to stray far from water. When caught short, elephants have even been known to draw water from their throats with their trunks and spray it over their ears. One way or another, an elephant's magnificent flaps represent a rather elegant solution to the problem of being a 5-tonne mammal with an overactive metabolism in a hot climate.

Both the present-day ecology and the evolutionary history of elephant-like creatures is consistent with this interpretation of their ears. Modern elephants, for example, are classified into African and Indian species, and the African ones are further divided into the familiar savanna or bush elephant from eastern Africa and the little-known forest elephant from the equatorial woodlands of central and western Africa. Indian and African forest elephants inhabit shaded woodland where temperatures tend to be much lower than on the savannas where bush elephants live. If the temperature-regulation theory of elephant ears is correct, then bush elephants should have larger ears (or at least ears capable of holding more blood, either by being larger or by having a more dense network of blood vessels). And sure enough, though Indian and African forest elephants have ears that are huge by normal mammalian standards, they are considerably smaller than the enormous flaps that grace the heads of their bush-dwelling cousins.

Extant elephants of all kinds live in relatively warm climates, but some extinct species of mammoth lived in very cold places indeed. We know most about the anatomy of mammoths from animals frozen in the tundra regions of Siberia. Fur and thick layers of fat are preserved in some specimens, a sure sign that mammoths were vulnerable to cold. But what about their ears? If mammoth ears were for sexual display or intimidating rivals, then it is possible that they could have been large but poorly supplied with blood, or covered with fur, or tucked in close to the body most of the time. Or perhaps heat loss through their ears could have been compensated for by thicker fur or an extra layer of fat. But if the ears of elephant-like creatures are primarily devices for letting out heat, then it would be very odd for those living in bitterly cold places to evolve thick layers of fur and fat yet have big, fleshy, blood-filled ears sticking out. And indeed, though the mammoths of Siberia were majestic animals in many ways, there is one anatomic department in which they were distinctly lacking compared with their living relatives: they had tiny, furry ears. It all fits together nicely. The magnificent ears of elephants are, quite literally, radiators.

While elephants are odd creatures in many respects, their bodies conform to a relatively simple set of design criteria. Their thick legs, stately carriage, short necks, trunks, heat production, bald skin, and big ears are all consequences, one way or another, of the inescapable scaling relationship between areas and volumes. Elephants make satisfying scientific subjects because the discovery of basic principles that explain diverse and seemingly unrelated phenomena is what science is all about. But the anatomy of elephants is just the beginning. The ways in which animals of different sizes and types consume and manipulate energy exerts a fundamental control over numerous patterns of life on our planet. We have figured out elephants and their ears. Next, shrews and their extraordinary hearts.

[2]

THE ROAD OF LIFE

Imagine a road 100 km (62 miles) long with a 5-tonne (U.S. 5.5 tons) elephant at one end, a fairy-fly weighing 0.000001 gram (.000000035 ounce) at the other, and representatives of the rest of Earth's creatures arrayed between in order of size. From a high vantage point, one pattern in our parade of animal life is immediately obvious: there are far fewer species at the elephant end than at the fly end. The elephant and a white rhinoceros, the two largest land animals on Earth, are separated by a full 50-km (31-mile) stretch of empty road. A little farther on from its wide-mouthed African cousin stands a one-horned Indian rhino. At 60 km (37 miles), standing only 1.4 meters (4.6 feet) at the shoulder but weighing in at around 2 tonnes stands a hippopotamus.[1] A black rhino and a giraffe stand at the 75-km (47-mile) mark. At 80 km (50 miles) (1 tonne) there is a water buffalo, a Javan rhino, and the first reptile, an estuarine crocodile. Two kilometers (1.2

---

1. There is considerable variation in weight among elephants, white rhinos, and hippos, so these positions along the road are approximate. African bush elephants usually weigh 4 to 7 tonnes (U.S. 4.4 to 7.7 tons), but large males (if they survive the poacher's gun) can reach 13 tonnes. African forest elephants, desert elephants (a race of the bush elephant), and Asian elephants are generally smaller, but the largest males may top 8 tonnes. The maximum weight for a white rhino is about 3.6 tonnes. Really big male hippos have been known to reach 4 tonnes.

miles) farther on stand a yak, an Asian gaur, a buffalo, a Sumatran rhino, and a male giant eland. (Gaurs are huge cattlelike animals from India and Southeast Asia. Elands are the largest members of the antelope clan of mammals, inhabiting open plains, savanna, montane forests, and semiarid regions of Africa.) The largest camel stands at 90 km (56 miles) alongside a polar bear, a Kodiak bear, and a moose (the largest deer). Grevy's zebra, the heaviest member of the horse family, stands at 92 km (57 miles). Between 92 km and 99.98 km (62 miles), all the remaining members of the deer, bear, camel, horse, pig, antelope, cat, and dog families crowd in, right down to the tiny fennec fox at 1 kg (2.2 pounds).

There are some interesting anatomic patterns among the mammals on this highly populated stretch of road. As body weight decreases, skeletons become relatively lighter, and the animals become more agile and acrobatic. The lighter runners have less mass and inertia, which means they can accelerate, decelerate, and corner more effectively. Lighter animals like squirrels move around with bent legs back *and* front, and are agile enough to climb vertically up tree trunks and jump from one tree to another. These patterns support the general idea that increasing body weight in the animal kingdom is counterbalanced partly by a much more circumspect approach to locomotion.

Fur thickness also varies among these mammals, but not in a simple, consistent way. The general trend is for insulation to thicken with increasing body weight, mainly because large animals can carry more fur without tripping over it. But the pattern can be distorted by environmental influences. Among African savanna antelopes the trend is reversed, with smaller antelopes being hairier than larger ones. This suggests that larger antelopes may have more of a problem keeping cool than warm. High temperatures on the savanna and the fact that antelopes periodically have to exert themselves to the limit fleeing from predators is the most likely explanation. The baldness of savanna rhinos and elephants, therefore, may just be an extension of the general trend toward hair loss among the larger inhabitants of this hot and exposed grassland environment.

The last 60 meters (66 yards) of road is very crowded indeed. Over 99 percent of the 1800 species of rodents squeeze into this stretch, most in the last 20 meters (66 feet). They are accompanied by most of the world's 9700 bird species, 3800 lizards, 3800 frogs and toads, 2700 snakes, and 1000 bats. Most of the 900 species of rats and mice squeeze into the last 2 meters (6.6 feet) at body weights below 100 grams (3.5 ounces). All 319 species of hummingbirds take up stations within the last 40 cm (16 inches). Below 20 grams (.7 ounce) the number of warm-blooded species declines, but most bats (around 820 species) and shrews (280 species) are packed into the last 30 cm (12 inches).

Four centimeters (1.6 inches) from the end of the road, three warm-blooded animals stand abreast as the smallest of their kind. Kitti's hog-nosed bats were discovered in 1973 roosting in the deepest chambers of some isolated limestone caves in Thailand. At 2 grams (.07 ounce) apiece, the global population of 2000 animals weighs substantially less than an elephant's ear. Half a world away in the forest of Cuba you may be lucky enough to see, if your eyesight is sharp enough, a little bird with iridescent blue feathers darting between the flowers of the forest floor. With a body length of just over 1 cm (.39 inch), the bee hummingbird is aptly named. And in grasslands throughout southern Europe, southern Asia, and Africa, the world between grass stems is patrolled by the tiny Etruscan shrew (Fig. 2.1). So small that they can crawl into the burrows of earthworms, these pocket titans are nevertheless among the most voracious predators on the planet.

Surrounding the last three warm-blooded tetrapods at the 2-gram mark are many cold-blooded ones. Reptiles and amphibians are very common at this weight, and many are markedly smaller.

FIGURE 2.1 Etruscan shrew (aka pygmy white-toothed shrew). Adult weight around 2 grams (.07 ounce); head-body length 4.5 cm (1.8 inches).

Two species of gecko from the islands of Virgin Gorda and Haiti, a chameleon from Madagascar, and several frogs from Cuba, Mexico, Argentina, the West Indies, and Brazil are all smaller than houseflies. Why do mammals and birds wink out of existence at such high body weights? The answer lies in the different ways in which warm-blooded and cold-blooded animals use energy and, once again, the scaling relationship between areas and volumes.

Elephants and Etruscan shrews may differ in weight by a factor of 2.5 million, but both are twigs on the mammalian family tree, and both endeavor to maintain their bodies at the standard mammalian operating temperature of 38°C. However, tiny animals have a large area of skin relative to the volume of their heat-producing tissues, which means that they heat up and cool down very quickly (which is why small African mammals in temperate zoos are kept in heated buildings, while the elephants are let outside). And the problem of heat loss is compounded by the limited amount of insulation that small animals can carry. Nearly all mammals are covered with hair, and the greater the problem of heat loss, the more effective this insulating layer needs to be. Arctic foxes regularly cope with temperatures down to minus 50°C (minus 58°F) and have evolved very thick pelts as a consequence, but if an Etruscan shrew had fur as thick as an Arctic fox it would look like a pom-pom and be about as ecologically viable. Not only does heat escape from small animals very easily, there is little they can do by way of insulation to stop it without immobilizing themselves.

Etruscan shrews cope with this problem partly by sheltering but mainly by what amounts to metabolic brute force. The extent to which these animals have stoked up their internal boilers to achieve thermal balance is quite remarkable. An adult elephant may spend most of its waking hours eating or searching for food and may consume 300 kg (660 pounds) of vegetation every day in the process, but this huge amount of herbage amounts to only 4 percent of its body weight. A shrew, on the other hand, consumes 130 percent of its body weight every day, and this is of animal tissue, which has a much higher nutritional value than vegetation.

And the rate at which shrews consume oxygen is even more staggering. Vertebrates of all kinds dissolve oxygen in blood and pump it through a network of blood vessels to the sites where heat-producing metabolic reactions occur. Our hearts beat at 60 to 80 beats per minute when we are at rest and two to three times this rate during exercise, but the heart of an Etruscan shrew races at an astonishing 1200 beats per minute. Even this would be woefully inadequate if shrews had hearts of normal mammalian proportions (about 0.6 percent of body weight). Hearts have to contract, expel blood, relax, and fill up again, and there are physical limits to the speed with which this cycle can be accomplished. A heart weighing 0.012 gram (.0004 ounce)—0.6 percent of 2 grams—would have to beat at over 3500 beats per minute to get enough oxygen around a shrew's body, or about three times faster than physically possible. As a heart cannot go much faster than 1200 beats per minute, the only other way to increase blood flow is to make it bigger, so that more blood is expelled with each stroke. The heart of an Etruscan shrew actually weighs 0.035 gram (.0012 ounce), or about three times the relative size of a thoroughbred racehorse's. Only with hearts of these heroic proportions can shrews tick over fast enough to maintain their bodies at a temperature of 38°C.

Etruscan shrews, Kitti's bats, and bee hummingbirds all have similar appetites and enormous hearts, and this is the fundamental reason why 2 grams is about as small as a warm-blooded animal can be.[2] A 1-gram (.035 ounce) shrew would have such a large amount of skin relative to its heat-producing tissues that it would require an impossible metabolic rate to maintain its body temperature. If such a diminutive animal had very effective insulation, it might just be able to pull off the feat of surviving, but what sort of insulation could a 1-gram animal carry and still move around? It seems reasonable to conclude that Etruscan shrews, Kitti's bats, and bee hummingbirds are at least pushing the boundaries of what is possible for warm-blooded animals.

And the same reasoning explains why warm-blooded animals only come in a limited variety of shapes. A spherical animal would have the minimum possible amount of skin relative to its internal tissues, and any departure from sphericity would necessarily increase the relative amount of surface. This is why there are no warm-blooded animals shaped like garter snakes. Extremely thin animals have very large surfaces relative to their volume, and for a warm-blooded animal this would mean a large area through which precious metabolic heat could escape. Weasels are about as

---

2. Two weeks after I had written this, Jon Bloch of the University of Michigan reported to the 1998 meeting of the Society for Vertebrate Palaeontology in Snowbird, Utah, the jawbone of a 50-million-year-old insectivorous mammal called *Batodonoides*. The length of the jaw (8 mm or .3 inch) and teeth (ca. 0.75 mm or .03 inch) suggest a body weight around 1.3 grams (.05 ounce), smaller than any living species. Reconstructing whole creatures from skeletal fragments involves some guesswork, so I will stick with 2 grams (.07 ounce) as the official lower limit, but *Batodonoides* and other extinct mammals may have been smaller.

snake-shaped as mammals get, an adaptation that has allowed them to earn a living hunting in awkward places like burrows. Skinniness has been a highly successful evolutionary innovation for these animals, but one that has had to be bought. Because of their elongated shape, weasels metabolize energy almost twice as fast as more portly mammals of the same weight.

Cold-blooded animals like lizards, some snakes, and caecilians (wormlike aquatic or burrowing amphibians) are the masters of being both small and skinny. As every herpetologist knows, lizards spend much of their time holed up in burrows, under stones, between rocks, in the rubble at the bottom of scree slopes, or under fallen logs. Cold-blooded ambush predators need to hide from the midday sun, the chill of night, and the attention of their prey and predators, so being small and thin has obvious advantages. Thin animals are also lighter than fat ones, which is why geckos can crawl up walls and across ceilings. Defying gravity has allowed this particular group of lizards to exploit refuges and hunting grounds that are off-limits to nearly all other types of tetrapod. (The main attribute that wall-walkers must have is low weight because then there is less for gravity to pull on. The toe pads of a gecko's foot are also finely divided into minute, flexible tufts that under high magnification look like tiny cauliflower florets. The toe pads have a large surface area that comes into very close contact with the wall. Molecular attraction is very strong over short distances, which allows geckos to stick to even the smoothest surfaces.)

So warm-blooded animals forgo many potential ways of life because they are so large and squat, and the enormous diversity of small cold-blooded creatures on Earth demonstrates that these opportunities are considerable. And this leads to an obvious question: if the advantages of being cold-blooded are so manifest, why haven't mammals and birds evolved some sort of lower-powered metabolic system? If a warm-blooded animal could reduce its energy demands in some way—adopt some effective energy-saving strategy—it could conceivably be smaller or less rotund without having to increase its food intake and cardiac output. If warm-bloods were smaller, they could then vie for the numerous fertile niches currently occupied exclusively by their cold-blooded cousins.

Energy-saving strategies are, in fact, quite common among warm-blooded animals under certain environmental circumstances. Some larger mammals, such as badgers, raccoons, skunks, and bears, undergo a period of winter lethargy when food is scarce, during which they reduce their body temperatures. The reduction is rarely dramatic—a drop from 38°C to 34°C is common in bears—so most of the energy saving can be attributed simply to inactivity. This type of winter lethargy is often called hibernation, but physiologists usually reserve this term for small animals with the ability to reduce their temperatures by as much as 30°C (54°F) and enter an extreme state of torpor. Hummingbirds, for instance, cannot feed at night, and small nocturnal bats cannot forage during the day, and many species of both types of animal frequently have to contend with seasonal or unpredictable variations in food supply. Under such circumstances, extreme hibernation can make the difference between life and death. Allen's hummingbirds consume about thirty times less oxygen when they are torpid at night than when they are awake, which adds up to a daily energy saving of around 30 percent.

So extreme hibernators demonstrate that warm-blooded animals can evolve a metabolic strategy similar to that of cold-blooded ones (becoming torpid is, after all, what cold-bloods do when they are at rest in a cool environment). Although hibernation is a very effective energy-saving technique and quite sophisticated in detail, it is rather crude in principle—analogous to a heater with a simple on-off switch. It is clearly not a viable route to smaller size or extreme skinniness because when bats and hummingbirds are active, they operate at full throttle with outsized hearts beating at close to the maximum theoretical rate. But as hibernators have the ability to alternate their body temperature between two extremes, what is to stop them from making similar energy savings by stopping somewhere in between? Why not operate at, say, 30°C (86°F)? This temperature could be maintained with a lower metabolic rate, less food, and a slower, smaller heart. In fact, why have a characteristic operating temperature at all? Why don't mammals and birds simply let their body temperatures vary with the environment and bank all the saved energy?

Well, there are some mammals that have evolved lower body temperatures and metabolic rates in response to certain environmental conditions. Insectivorous bats, for example, often have lower metabolic rates than nectar- or fruit-eaters. Flying insects tend to fluctuate in abundance more than fruit and flowers, so the low metabolic rates of insect-eating bats have probably evolved to see them through lean periods when food becomes scarce. Among tree-living herbivorous mammals, metabolic rates tend to decrease as the proportion of leaves in an animal's diet increases. It is more difficult to extract energy from tough, fibrous, often poisonous leaves than from fleshy, nutritious fruit, so the low metabolic rates of leaf specialists may be necessary to survive on a troublesome food resource. Some desert rodents also have low metabolic rates, especially if they feed on dry foodstuffs like seeds. A high metabolic rate is accompanied by a high demand for oxygen, rapid breathing, and thus the potential for a high rate of evaporation from the lungs, so the low metabolic rates of desert seed-eaters may be an adaptation to limit water loss. These examples show that warm-blooded animals have the capacity to downsize their metabolic engines when it is beneficial to do so. None of the animals mentioned so far, however, could be called cold-blooded in the extreme sense that lizards and frogs are. But there is one animal that demonstrates unequivocally that warm-blooded animals do have the evolutionary potential to shake off their metabolic

FIGURE 2.2 The naked mole-rat, perhaps the most curious mammal on Earth. Head-body length 8 cm (3.1 inches).

heritage almost completely. To find this surpassingly strange creature we must leave the sunlit surface of the Earth and enter the perpetual darkness of the subterranean world.

Beneath the arid grasslands of Somalia, Kenya, and Ethiopia lives an animal that heads the list in nearly every category of mammalian oddness. Naked mole-rats are hairless, colonial, burrowing rodents that live and breed like termites. They have loose, fleshy skin hanging in folds around their necks and bodies, and huge, exposed upper and lower incisors like the grabbers of a mechanical digger (Fig. 2.2). If you are squeamish about rodents, give naked mole-rats a miss. Even more remarkable than their appearance is their social system. A colony consists of a single female or queen who does all the breeding, and a sterile caste of small-bodied workers who forage for food and excavate the extensive system of burrows. And just to emphasize the estrangement of these creatures from the rest of the *Mammalia* in the strongest possible terms, the internal temperature of naked mole-rats is precisely determined by the temperature of the air in their burrows. In other words, they are essentially cold-blooded. (Technically, poikilothermic: they do not regulate their temperature by metabolic or behavioral means.)

This radical departure from standard mammalian design has given rise to much speculation about the environmental conditions under which naked mole-rats evolved. The climate underground is much more stable than at the surface, where sun, wind, rain, and frost cause temperatures to fluctuate widely, and the climate of sub-Saharan Africa is relatively warm and stable all year round anyway, so a subterranean species freed from the need to respond physiologically to such changes would appear to be in a good position to do away with thermoregulation altogether. However, there are eight species of burrowing mole-rat in sub-Saharan Africa, and the other seven have at least some control over their body temperatures, so a subterranean existence per se cannot be the whole solution.

Might the answer be connected with the unique social system of naked mole-rats? In situations where only one female in a colony breeds, some unusual evolutionary innovations can arise. The process of natural selection affects the frequency of genes in a population through the effects that they have on individual animals. The genes of individuals who are best equipped to survive and reproduce will tend to be better represented in future generations. But the situation is more complex for animals that live in cooperative colonies with a single breeding female. How can worker females further their own genetic ends when they do not give birth to any offspring? The only viable option for sterile females is to ensure that their queen survives and is successful at reproducing. This roundabout approach works because nonbreeding females are more closely related to their own queen than to any other. If the colony as a whole is successful, it is more likely to give rise to new colonies, so the genes of individuals in successful colonies, whether they breed or not, will march on into the future. Under the curious social system of naked mole-rats, therefore, the good of the individual corresponds closely with that of the colony, so we might expect some adaptations that benefit the whole collective to be favored by natural selection. Cold-bloodedness is probably one such adaptation, but to understand why, we have to look in more detail at the environment in which these animals live and how sociality improves their chances of survival.

Differences in the size and social system of mole-rats seem to be linked to differences in aridity. As ground conditions become drier, colony size and the level of cooperation between colony members tend to increase, while the size of individual mole-rats tends to decrease. The naked species lives in the driest areas, has the most highly developed social system, and the smallest individuals. It seems to be the intermediate effect of aridity on plants that underlies these patterns. Mole-rats eat mainly roots and tubers, and as ground conditions become drier, tubers become larger and more widely spaced. Large tubers contain a lot of food and water and do not desiccate easily, so the production of a few big tubers, rather than lots of little ones, is a useful adaptation for plants that have to contend with drought. Burrowing uses up large amounts of energy—probably 3500 times the amount required to walk the same distance aboveground—so mole-rats in arid regions run a high risk of starvation because bumping into large tubers is unlikely. In arid areas, therefore, provided food is shared among colony members when found, it would make sense to have more individuals per colony burrowing in all directions to maximize the chances of finding food. But simply increasing the number of individuals would be a risky strategy because the total food demand of the colony would be higher, and this would increase the pressure on tubers that are in short supply anyway. The only way of having more burrowers for the same total food demand is to make each colony member smaller. The small body size of individual naked mole-rats, therefore, is at least consistent with the idea that survival of the colony as a whole is of paramount importance.

However, while it is true that a hundred small mammals consume less energy than a hundred large ones, smaller mammals normally incur a substantial metabolic cost because they have higher metabolic rates per unit of body tissue (remember that Etruscan shrews consume 130 percent of their body weight in food each day, compared with only 4 percent for elephants). But size reduction in the naked species has been achieved without this cost because colony members do not expend energy to keep themselves warmer or cooler than their surroundings. So cold-bloodedness and small body size in naked mole-rats seem to make sense. The low energy demand of each individual adds up to a low total demand for the colony as a whole, and a hundred small mole-rats burrowing in all directions stand a better chance of bumping into widely scattered tubers than ten big ones.

I have deliberately sidestepped the issue of the order in which these changes in mole-rat ecology and physiology came about. Cold-bloodedness, increased sociality, and small body size may all have evolved in parallel in response to increasing aridity of the environment, or each characteristic may have evolved at a different time, and perhaps for different reasons. Whatever the detailed sequence of events, naked mole-rats have clearly gone one step further toward a fully cold-blooded lifestyle than bears, hummingbirds, and insectivorous bats. Their energy-saving strategy is neither an on-off affair nor a controlled suppression of metabolic rate, but a complete abandonment of temperature regulation altogether. Mole-rats demonstrate that warm-blooded animals have the ability to evolve into functionally cold-blooded ones.

If warm-blooded animals are ever to breach the 2-gram (.07-ounce) glass floor that so restricts their evolutionary potential, then, in metabolic terms at least, naked mole-rats are clearly the front-runners. With an average body weight of 28 grams (1 ounce) they have a long way to go, and their demanding excavatory lifestyle may well militate against their getting any smaller at all, but neither can we assume that these creatures have reached the end of their particular evolutionary trajectory. If a hundred small mole-rats burrowing in all directions have a greater chance of encountering widely scattered tubers than fifty larger ones, then perhaps two hundred *really* small ones would be even more successful. And now that they have shaken off the constraints of temperature regulation altogether, what other ecological opportunities might they have the potential to exploit? Will the descendants of naked mole-rats one day emerge into the sunlight from their subterranean birthplace and begin to take on other cold-bloods at their micro-ecological game? Evolution hasn't finished yet, and stranger things have happened.

Perhaps the biggest puzzle is why warm-blooded animals have not traveled the evolutionary road toward cold-bloodedness more often. After all, maintaining a high body temperature uses up an enormous amount of energy that could be used for other purposes regardless of the type of environment in which an animal lives. Why are cold-blooded mammals and birds so rare? At present there is no definitive answer to this question, but a number of factors are probably involved.

Naked mole-rats notwithstanding, the Earth today is overrun by four-legged animals that are clearly either warm-blooded or cold-blooded. Tepid animals must have existed when the transitions from cold-blooded to warm-blooded occurred, and animals with a range of body temperatures may have lived side by side in the distant past, but most tetrapods alive today fall rather obviously into one group or the other. For mammals this relative uniformity may be due to the fact that for the first two-thirds of our evolutionary history we were small, insectivorous, and probably nocturnal. Warm-bloodedness has obvious advantages for animals that come out at night, so mammals would have been ideally suited to this particular niche. But when a group of animals spends 140 million years dedicated to a particular way of life, it may be difficult to throw off the shackles of history. A good example of this phenomenon is color-blindness. The incidence of this condition is very high among mammals, and many researchers believe it to be a hangover from our long history of nocturnality (color vision being of no use at night). Some mammalian groups have since developed color perception—primates, including humans, being an obvious example—but the trend has been rather limited. A long period of enforced warm-bloodedness because of the demands of a nocturnal lifestyle may similarly have limited the capacity of mammals to alter their metabolic characteristics.[3] Even if valid, this argument will not work for birds because birds have always been predominantly daytime animals, but high muscle temperatures are required to sustain the extremely energetic demands of flight, so the retention of full-blown warm-bloodedness in this group of tetrapods is perhaps less surprising.

There are also a number of ecological reasons why the transition from warm-bloodedness to cold-bloodedness may not be easy. Every step along this particular evolutionary road would be subject to rigorous testing in the ecological arena. At its end, a hypothetical cold-blooded shrew would have to compete with all the lizards, snakes, and frogs that have been exploiting this sort of lifestyle successfully for millions of years; but it may not be easy to travel any distance along the road at all. Evolution happens in small steps, so the most rigorous test for an animal setting off toward cold-bloodedness would come from closely related animals competing for the same food, burrows, mates, nesting materials, and so on. In other words, a low-powered shrew would initially

---

3. Although no fundamental biological barriers to the transition as a result of our nocturnal history have to date been identified. In fact, no one has put forward a convincing defense of the idea that the transition from warm-bloodedness to functional cold-bloodedness should even be difficult at the cellular or subcellular level, and the argument that there must be some sort of resistance to the transition because cold-blooded mammals and birds are so rare is obviously circular. Provided we do not require a mammal to evolve a metabolic system exactly like that of a lizard or frog in all cellular and biochemical details—which is just highly unlikely—then there are at present no empirically supportable arguments to defend the view that the evolutionary transition from a mammal-like to a lizardlike metabolic regime is hampered by any fundamental biological barriers.

have to compete with the full-powered variety. It might benefit in some ways by having lower energy requirements, yet it might lose out in competition for food (because the high-powered versions are more active foragers), come off second best in active disputes over burrows or mates, or be an easier target for predators. If a mutant shrew adopted a body temperature of 28°C (82°F), it would be more susceptible to heat stress, because the temperature of the environment might often exceed 28°C, although seldom the far higher body temperature of a normal shrew. A shrew with a low body temperature would also have a slower rate of digestion and a slower rate of chemical reactions within its cells. Mammals with low metabolic rates also tend to have longer intervals between generations and smaller litters of offspring, so high-powered shrews might be able to outcompete low-powered variants simply by outbreeding them. All in all, it might be very difficult for a low-powered shrew to get far enough down the evolutionary road to free itself from the competitive sphere of high-powered shrews, mice, rats, and predators that are partial to small furry things, particularly if they are slower than usual. Only when the benefits of a reduced metabolic rate outweigh all the disadvantages is a trend toward metabolic downsizing likely to be favored by natural selection, and the areas of the ecological stage where this applies to mammals and birds might be rather restricted.

Whatever the exact mechanism that constrains mammals and birds in their metabolic profligacy and thus their particular range of sizes and shapes, the last 4 cm (1.6 inches) of our 100-km (62-mile) road is completely free of high-powered animals. This may not seem much of a missed opportunity, given the spectacular diversity of warm-blooded animals on the first 99.99996 km of road, but the fact is that 99.9 percent or more of all animal species on Earth cram into this last short stretch. Just among the insect clan of animals, over 1 million species have already been named and described, and 10,000 new species are discovered annually.

By some estimates there may be 30 to 40 million species of insect on Earth; add to this 70,000 arachnids (spiders, scorpions, harvestmen, mites, and ticks), 65,000 worms, 60,000 molluscs, 42,000 crustaceans, and 11,000 myriapods (millipedes and centipedes) and the last 4 cm of tarmac become a towering heap of species stretching miles into the sky.

The ubiquitous presence of insects in this mountain of microspecies raises some interesting questions, because most insects routinely engage in the extremely strenuous activity of flight. We have already seen that small mammals and birds are limited in the extent to which they can increase their metabolic rates because of the problem of supplying enough oxygen to their tissues. But sphinx moths fly at body weights of 1 gram (.035-ounce), and they have to prewarm their flight muscles to 35°C (95°F) by a process analogous to shivering before they can even get into the air. Heat escapes from a 1-gram sphinx moth even faster than from a 2-gram shrew, so how do sphinx moths acquire enough oxygen to elevate the temperature of their diminutive bodies to such high levels? The answer lies in the radically different respiratory system possessed by insects. Vertebrates dissolve oxygen in blood and pump it around their bodies in a network of blood vessels. Oxygen has to diffuse into the blood in the first place, and it then takes a long time to get it to where it is needed—around four seconds for a complete circuit of a shrew, and two and a half minutes for an elephant. In contrast, insects pump air around their bodies in tubes from where oxygen diffuses straight into their tissues. This arrangement is simply more efficient at distributing oxygen and allows insects to fly at very small body sizes.

By regulating the heat generated by their flight muscles, some moths can fly with body temperatures maintained at 35°C to 40°C (95°F to 104°F) at body weights down to 0.2 gram (.007 ounce), ten times lighter than the smallest bats and birds. But as insects approach this size, storing heat becomes increasingly difficult because it escapes so easily into the environment. Temperature regulation becomes practically impossible at weights below 0.2 gram, so insects fly with muscle temperatures close to that of their surroundings. Fairy-flies and feather-winged beetles at one-millionth of a gram (.000000035 ounce) necessarily remain at the temperature of the environment even in flight, but for these nano-creatures flying is less of a problem than staying on the ground, as the slightest breath of wind can carry them high into the air along with pollen and dust and the rest of the aerial plankton that constantly sails around the globe.

With fairy-flies we have reached the end of our animal parade. A hundred kilometers might seem like an odd or arbitrary distance to have chosen, but the road had to be this long to accommodate an Etruscan shrew in the last 4 cm (1.6 inches) (even so, it would have to hunch up a bit and curl its tail around). I have also been selective not only with the animals we have met en route but also with those at either end. Blue whales are actually twenty times heavier than elephants, and the bacteria-like organism *Mycoplasm* weighs less than 0.0000000000001 gram (0.0000000000000035 ounce), but including these extremes would have made the road, and the narrative, unreasonably long.

The creatures we have met in the last two chapters illustrate some of the ways in which different types of animals fuel their lives and how their energy demands vary with body size. The ecological, biogeographic, and evolutionary ramifications of these fundamental differences are the subject of the rest of this book.

Our jaunt along the road of life has highlighted one characteristic of birds and mammals in particular. The members of these two great classes alone are capable of keeping themselves warm solely by their own metabolic exertions. Lizards may be as warm inside as rats for most of the day, but they achieve this thermal constancy with a low-powered metabolic engine and heat absorbed from the environment. Evolution has tinkered with the otherwise standard-issue mammalian engines of insectivorous bats and naked mole-rats to suit them to their peculiar environmental circumstances. Warm-blooded hibernators have a rev-limiter that allows their engines, when required, to tick over slowly without stalling. Pythons sometimes warm themselves by rhythmically contracting their skeletal muscles. Even some plants, such as North American skunk cabbages, manage to store some of their metabolic heat and thereby remain slightly warmer than their surroundings. All these unusual cases are fascinating, but no biologist would confuse the functional warm-bloodedness of pythons and sphinx moths with the high-powered metabolic engines characteristic of mammals and birds. In warm-blooded animals, evolution produced something altogether different.

Time to bite the bullet. How did this great metabolic divide come about?

[3]

LIFE HOTS UP

Four hundred million years ago the world would have been an eerily unfamiliar place to human eyes. Only invertebrates such as centipedes, millipedes, scorpions, springtails, mites, and spiders inhabited the vast expanses of dry land around the globe, and the only backboned animals around were fish. By the Carboniferous period (Fig. 3.1)—so called because large areas of Britain at this time became covered by vegetation destined to become coal—four-legged animals were making their way onto land. Amphibian-like creatures were the first to exploit the rich pickings on offer beyond the water's edge, although they remained tied to aquatic habitats, as all amphibians are today, for the purposes of breeding. By the middle of the Carboniferous, animals with the knack of laying eggs on dry land began to appear. These creatures diversified rapidly into groups that formed the ancestral stocks for all later reptiles, mammals, and birds. Synapsid reptiles—those with one opening in the skull behind each eye socket—eventually gave rise to mammals. Anapsids—with no openings behind their eyes—gave rise to turtles and tortoises. And two branches of the diapsid lineage—two openings—led to lizards and snakes on the one hand, and crocodilians and birds on the other.

The consensus among paleontologists is that the last common ancestor of both mammals and birds was a cold-blooded reptilian

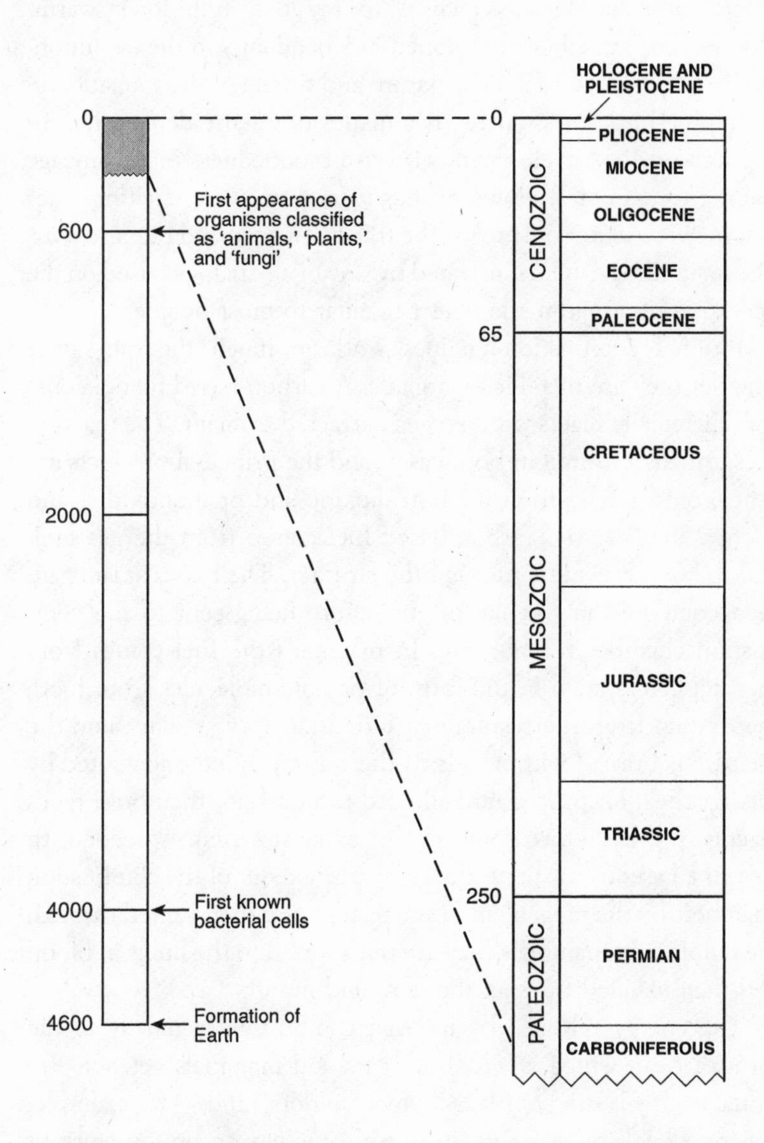

FIGURE 3.1 The geologic timescale. Numbers are millions of years before the present. The right-hand column is an exploded view of the most recent periods in Earth's history.

animal of some kind, which is to say that full-blown warm-bloodedness must have developed independently in the evolutionary lines leading to each. The nature and timing of the transition in the reptile-bird line is currently a matter of intense debate. Recent research on the development of warm-bloodedness in the lineage leading to mammals, however, has turned up some exciting clues that may reveal not only *when* the transition occurred but also *why*. The main ideas can be illustrated by way of an analogy based on the operation of a man-made object familiar to most people.

Internal combustion engines work in much the same way whether they are metallic or organic. A carbon-based fuel of some sort chemically reacts with oxygen extracted from air. The reaction is essentially burning in both cases, and the principal products are water, oxides of carbon (carbon dioxide and/or monoxide), and energy. In a car the carbon-based fuel comes from the gas tank and the oxygen enters through the air filter. The two reactants are then combined in the carburetor before being sent to the combustion chamber for burning. In mammals the fuel comes from the digestive system in the form of organic molecules from food, the oxygen from the respiratory and circulatory systems, and the mixture is burned within cells. If the waste products generated by this combustion process are allowed to build up, then both types of engine will stall, so some sort of exhaust system is needed. In a car the gaseous waste products are pushed out of the combustion chambers by the movement of the pistons and then vented through the tailpipe. In mammals they are transported to the lungs in blood and then exhaled through the nose and mouth.

The energy released by internal combustion is mostly in the form of heat, which is why both cars and mammals get hot. Residual energy is used to fuel activity of various kinds—for example, driving wheels or legs—or, in organic engines, to do the work of manufacturing useful substances. In animals, one type of molecule manufactured in cells, called adenosine triphosphate, is particularly important because it contains high-energy bonds between some of its atoms. These bonds are essentially power-storage units, like batteries, and they can be broken open whenever free energy is required for some biochemical purpose or to fuel the muscle contractions required for moving around.

Both mechanical and organic engines need a constant supply of oxygen to sustain the reactions at the heart of the combustion process, so we can call them "oxygen-based" or "aspirated." Assuming that fuel is available in copious amounts, the speed at which an aspirated engine can turn over depends on the maximum rate at which oxygen can be supplied. If engine turnover exceeds this rate, the oxygen will run out and the result, again, will be a stall. One way around this problem is to have a backup system that does not require oxygen, so that when the maximum capacity of the aspirated engine is exceeded, the non-oxygen-based engine can cut in. We will call this backup system a supercharger. In a car the supercharger might consist of two small tanks of fuel that deliver an extra boost of power to the pistons when the contents are reacted together. The downside of using a supercharger is that particularly stubborn and noxious waste products accumulate much faster than they can be vented through the exhaust system. This buildup inhibits the flow of fuel, causing the supercharger to cut out. Moreover, these troublesome substances take so long to flush out of the engine once they build up that wise drivers reserve their superchargers for emergency situations and even then use them only for a few seconds at a time.

Animals have a similar type of engine setup with the same benefits and drawbacks. An animal's aspirated engine operates sustainably only if it ticks over slowly enough to stay within the limits of its oxygen-supply system. There are other types of metabolic reaction, however, that can deliver energy to drive the contraction of muscles but do not require oxygen. The downside of using an organic supercharger is that a substance called lactic acid is produced that causes muscle fatigue, and is very difficult to flush out of cells once it accumulates.

Runners in 100-meter races use their superchargers almost exclusively, because the effort involved in sprinting greatly exceeds the capacity of our aspirated engines. Sprinters get away with thrashing their superchargers like this because the job is done in a few seconds, after which they can go and sit down. In contrast, marathon runners use their aspirated engines almost exclusively, because they need to run at their maximum *sustainable* speed. The relative speed of marathon runners and sprinters over the ground gives a good feel for the enormous difference between sustainable activity powered by our aspirated engines, and unsustainable burst activity powered by our superchargers.[1]

How do the metabolic systems of cold-blooded and warm-blooded animals fit in with the engine analogy? Imagine a lizard and a rat of the same size sitting under a tree. If the lizard's aspirated engine is idling at, say, 100 rpm, then the rat's will be running at around 1000 rpm. Idling speed is the equivalent of an animal's resting metabolic rate, the basic physiological output necessary to keep us alive and functioning, and is typically ten times higher in mammals than in reptiles of similar size. A turnover of 100 rpm is very slow. If a metallic engine idled at this speed, the motion of the pistons and the flywheel would be clearly visible and the number of engine cycles could be counted pretty accurately. The engine would

1. Oxygen-based and non-oxygen-based metabolic reactions drive the activity of different types of muscles. Endurance muscles driven by the aspirated engine have to be supplied with oxygen-rich blood, which usually renders them reddish in appearance. Muscles driven by the supercharger, in contrast, need much less oxygen and are generally white. Chicken legs are red because they use them all the time for running around, but their breasts are white because chickens fly only in short bursts, usually as an escape tactic. Pigeon breasts, on the other hand, are deep red because they fly regularly. A fillet of fish consists mostly of white muscle, but there are often red stripes running along each flank close to the skin. The red muscle is used for sustained swimming and the white for short bursts of speed.

produce some heat, but it would dissipate so readily into the surrounding air that the stationary parts of the engine could be touched without any risk of burned fingers. A lizard's capacity to produce heat is similarly low, and the lack of an effective insulating covering means that it escapes into the environment very easily. So unless a lizard can soak up the sun's energy in some way or elevate its temperature by exercising, its body will be almost as cool as the surrounding environment.

The rat's engine, however, is a different beast altogether. A thousand rpm is about the idling speed of a small car. No human observer could count the revolutions of a flywheel going this fast, and any sensible person would test the temperature of any stationary parts gingerly with a finger before grabbing hold. Fast engines get hotter than slow engines of the same size, and it is the high tick-over speed of a rat's engine, aided by the heat-trapping effect of its fur, that keeps it warm.

At the subcellular level, engine speed equates to the rate of heat-producing metabolic reactions within cells. The sites where these reactions take place are small oval structures called mitochondria, which are either more densely packed in rat tissue (in organs such as the heart, liver, and brain) or souped up with special enzymes—molecules that encourage chemical reactions to occur—to push metabolic processes at an accelerated rate (in muscle tissue). As we shall see, a high engine speed is useful for a number of physiological and ecological reasons, but it doesn't come free: the rat has to consume ten times more fuel and oxygen than the lizard just to accomplish the feat of doing nothing. This basic difference in maintenance expenditure is of crucial significance, and in later chapters it will help to explain some of the most fundamental ecological and biogeographic patterns on Earth.

Let us now see how the engines perform when pushed. The lizard gets up and decides to run to another tree a kilometer (.6 mile) away. Continuous locomotion over such a distance easily qualifies as sustained activity, so it has to be driven by the lizard's aspirated engine if it wants to get there without stopping. The maximum turnover that the lizard's engine can maintain during exercise without exceeding the capacity of its oxygen-supply system is about 1000 rpm, or ten times higher than its idling speed. As the rat sits calmly watching the lizard jog off into the distance, both of their engines are now running at about the same speed.

The rat decides to follow at full sustainable tilt. The distance to be covered means that its aspirated engine will have to do all the work just as the lizard's did. The maximum turnover that the rat's engine can maintain without running short of oxygen is about 10,000 rpm, which is also ten times higher than its idling speed. But because the rat's maximum rate of turnover is ten times higher than the lizard's, its sustainable speed across the ground is correspondingly greater. For 100-gram (3.5-ounce) rats and lizards at a temperature of 38°C sustainable speeds are about 88 meters (96 yards) and 13 meters (14 yards) per minute respectively.

After their run, and after a little basking by the lizard, both animals are once again resting comfortably under a tree with their bodies at 38°C. Suddenly there is a movement in the grass a few meters away. Rustling grass often indicates something edible, and the senses of both animals immediately perk up. At the same moment the rat and the lizard rush toward the indiscreet rustler. Who gets there first? Most people would intuitively bet on the rat, but the situation is not at all straightforward. This is not a sustained run over a kilometer (.62 mile) but a brief dash. Short sprints are supported by power delivered from the supercharger, and the most powerful one belongs, in fact, to the lizard: the rat's can increase its speed by a factor of two or so, but the lizard's can propel it across the ground up to twenty times faster than normal. The upshot is that the maximum *un*sustainable running speed for mammals and reptiles of the same size is not significantly different.

For 100-gram individuals at 38°C it is about 240 meters (262 yards) per minute, so the race to the rustler would be rather difficult to call.

These fundamental differences in engine setup and performance have to be accommodated by any theory purporting to explain the evolution of warm-bloodedness, because at some point in the distant past natural selection somehow transformed animals with lizard-style engines into animals with rat-style ones. Moreover, natural selection is a step-by-step process that progresses in a certain direction only so long as there are net benefits to be had, so a plausible theory must also explain how the former could have been transformed into the latter in a series of small steps with each yielding some net benefit. This is a crucially important point so it is worth illustrating with an analogy.

Imagine that cold-bloodedness and warm-bloodedness are two villages, C and W, connected by a number of paths. The only paths that natural selection can follow from C to W are those that run consistently uphill (where "up" represents "fitter" individuals). A path that climbs to the top of an intervening hill and then plunges down to W on the other side cannot be fully traversed because progress will grind to a halt on the hilltop where the fittest individuals are. Paths that plunge down from C into a valley before climbing up to W cannot be traversed because individuals in village C are fitter than any of those on the descending leg of the path (nor can any progress be made from the other end). This condition of the constantly ascending path is a very stringent one, and evolutionary theories often fall afoul of it. In particular, the condition prevents scientists from dwelling on the obvious advantages of final states—like having wings, for instance—and forces us to offer a plausible explanation for the advantages of wings *in all stages of development,* even when they would have been too small and poorly feathered for flight. For judging the plausibility of the various theories that have been proposed to explain the evolution of warm-bloodedness, the constantly ascending path will prove to be an invaluable image.

So what are these theories? The most intuitively obvious one simply states that powerful metabolic engines evolved to keep their owners warm, because being warm is generally a good thing. The group of animals known as monotremes, comprising echidnas and the duck-billed platypus, keep their bodies between 30°C and 32°C (86°F and 90°F), marsupials such as kangaroos between 34°C and 36°C (93°F and 97°F), placental mammals and ratite birds (ostriches, rheas, kiwis, and emus) between 36°C and 39°C (99°F and 102°F), and passerines (songbirds) between 40°C and 42°C (104°F and 108°F). The preferred body temperature of most active lizards, snakes, and large flying insects is typically in the mid- to high 30s (95°F to 102°F) too. So warmth seems to be a universally desired commodity, and there are a number of reasons why this should be so.

Firstly, most physiological processes go faster at higher temperatures. A useful rule of thumb is that a 10°C (18°F) rise in temperature doubles the rate of a process, so that, for example, digestion proceeds at roughly twice the rate inside a lizard at 38°C than at 28°C. If an animal can turn food into growth, offspring, and energy to fuel activity faster than its competitors, then it may have an edge in the struggle for life. Secondly, muscles contract faster and more powerfully at high temperatures, which has obvious effects on locomotion. This phenomenon is particularly striking in animals small enough to be almost instantaneously affected by changes in temperature. Ants, for instance, speed up and slow down visibly as they walk between patches of sunlight and shade, and they can be made to walk at specific speeds experimentally by varying the temperature of their walkways. Animals with greater control over their locomotion can move around as slowly as they like, but being warm gives them the option to be highly active if occasion demands. Thirdly, a high body temperature allows mammals and birds to avoid the potentially hazardous activity of evaporative cooling most of the time: if warm-blooded animals stayed at 28°C rather than 38°C then the temperature of the environment would exceed that of their bodies more often, and this would require more frequent episodes of sweating or panting.[2]

And the hot-is-good theory for the evolution of warm-bloodedness has an obvious extension. A high body temperature is clearly beneficial for a number of different reasons, so surely it would pay an animal to remain warm at all times. Anyone who has tried to start a recalcitrant car in winter or had one overheat in a traffic jam will know that engines have a relatively narrow temperature range within which they work best. Car engines are fitted with chokes to encourage them to warm up by their own exertions, and various cooling systems (circulating water, fans) to prevent them from overheating. With these control mechanisms car engines can cope with Norwegian winters, Spanish summers, rain, snow, day, and night without ever breaching their temperature tolerances. Similarly, mammals produce enough metabolic heat to warm themselves to 38°C but they also have a liquid circulation system that can be adjusted to divert heat to and from the skin, a shivering response to warm up, and often an evaporative cooling response to cool down. With effective insulation, mammals too can operate at peak efficiency at all times in virtually all earthly environments.

Animals with constant body temperatures, regardless of what the actual set point is, also reap a number of benefits at the biochemical level. Enzymes vary in chemical composition and structure, but every enzyme has a relatively narrow temperature range within which it works best. Enzymes can be designed to work best at 10°C (50°F) or at 35°C (95°F), but no molecule can be expected

2. Actually, only mammals above a certain size sweat or pant. Small animals have large surface areas relative to the volume of water inside them, so evaporative cooling would quickly lead to fatal dehydration. Like cold-blooded animals, small warm-bloods usually cool themselves by sheltering.

to work at its best over this entire range. The internal temperature of a cold-blooded animal living in a changeable climate may well vary over this sort of range, which means that its enzymes will speed up and slow down. A body maintained at a constant temperature, however, can house suites of enzymes that work at peak efficiency twenty-four hours a day. And delicate enzymes that would otherwise fall apart at high or low temperatures can be mollycoddled inside a warm-blooded animal's body to deliver their specialized benefits.

Taken together, so the theory goes, the advantages of a high and constant body temperature might have yielded a number of ecological and biogeographic benefits that could account for the evolution of warm-bloodedness. Animals with constantly warm bodies could have exploited the environment at night when cold-bloods were hiding. By wrapping themselves in effective insulation they could have invaded parts of the globe that were far too cold for animals with lower-powered engines. By evaporating water from their bodies they could have remained cool and active in hot environments, thereby outcompeting all the cold-blooded animals shading themselves under rocks. Animals with controllable on-board heating systems would not have had to hide when temperatures soared or plummeted, so they would not have needed to be small or crack-shaped, either. They could have been large and cubic and completely exposed to the elements if being so offered a viable existence somewhere in the environment. Plants and invertebrates had been exploiting extreme environments and nocturnal niches for millions of years before warm-bloodedness evolved, so the larder was fully stocked and ready to harvest. With only cold-blooded animals around, proto-warm-bloods would have faced little competition for any of these potential ways of life, and increasingly high and stable body temperatures might have been favored by natural selection until full-blown warm-bloodedness was finally achieved.

The hot-is-good explanation for the evolution of warm-bloodedness seems eminently plausible. In fact, the advantages of a high and constant body temperature in some ecological situations are so obvious that the idea is almost irresistible. There must have been selection for stable body temperatures at some point in the evolutionary history of mammals and birds—the elaborate control systems present in living species would not make sense otherwise. But our main concern here is with origins, and there are a couple of problems that cast serious doubt on the whole idea that warm-bloodedness evolved solely for the purpose of keeping animals warm.

First, the benefits of any particular adaptation are irrelevant if they are outweighed by the costs, and warm-blooded engines are very expensive to run. Mammals have to procure the extra food and oxygen required to fuel their engines, but they also have to build and maintain outsized fuel-supply systems (hearts, lungs, thick-walled arteries) as well as growing and regrowing an insulating covering of some kind. All this extra construction and maintenance work has to be paid for with energy that might otherwise be used for important activities like producing babies. Warm-blooded animals also breathe faster than cold-blooded ones of the same size, which can cause considerable problems. Lungs are moist structures exposed to a stream of air that is usually relatively dry. Moisture flows from wet to dry just as heat flows from hot to cold, so hyperventilating warm-bloods run a much greater risk of drying out than their more sedate cold-blooded cousins. These pluses and minuses mean that the costs and benefits of temperature regulation in early tetrapods are exceedingly difficult to assess, and if the calculations could be done at all they would undoubtedly yield different outcomes under different environmental circumstances.

The second problem with the hot-is-good theory revolves around the condition of the constantly ascending path. Recall that for natural selection to push an animal design from here to there, the intervening distance must be covered in small steps with each yielding a net benefit in terms of an animal's ability to survive and populate the environment with its offspring. This means that the thermal benefits of a slight initial increase in metabolic rate must outweigh all the associated costs, and this is the point at which most researchers begin to have serious doubts. A small increase in metabolic rate would either require an increased intake of food or diversion of resources away from other activities like growth and reproduction. The costs are obvious, but what would the thermoregulatory benefits be? Few, if any, seems to be the consensus opinion. The resting metabolic rates of mammals and birds are six to ten times higher than those of similar-sized reptiles at the same body temperature, and the difference is even greater when the temperature of the environment drops and the metabolic rates of reptiles follow. A mere doubling of metabolic heat production in a lizard would not turn it into a warm-blooded animal—nowhere near—but the cost in terms of energy expenditure would double. It is very difficult to see how natural selection could possibly have favored elevated metabolic rates when there were no net benefits to be had.

What if some hypothetical small lizard had very effective insulation? Could a doubling of its metabolic rate then produce enough heat to maintain a body temperature close to that of a mammal? This is possible, at least in principle, but the insulation would have to be exceptionally efficient, and efficient insulation is a double-edged sword. Fur and feathers stop heat escaping, but they also prevent it from getting in. A lizard with a fur coat, therefore, would be largely disenfranchised from its solar energy source, which would hardly put it at an advantage in competition with its naked peers.

So the hot-is-good idea seems to founder. In the initial stages of the warming-up process, the benefits of small increases in metabolic rate do not seem to outweigh the costs. Can the theory be

FIGURE 3.2 The Permian pelycosaur *Dimetrodon*. Length around 3.5 meters (11.5 feet).

made to work? There is, perhaps, one way. In 1978 Brian McNab, a biologist blessed with more than his fair share of good ideas, came up with an ingenious spin on the hot-is-good idea. He noted that the vast majority of mammals extant during the time of the dinosaurs were no bigger than guinea pigs, much smaller than most of their "mammal-like reptile"[3] ancestors (Figs. 3.2 and 3.3). A trend toward smaller bodies in the lineage leading from mammal-like reptiles to mammals had long been appreciated, but it was McNab who saw the possible implications for the evolution of warm-bloodedness.

As already noted, large animals warm up and cool down more slowly than small ones because they have relatively small surfaces. If a cold-blooded animal were large enough, it could get through the cool of night with little change in its body temperature. If it also lived in the tropics, where temperatures are generally high and seasonal changes minimal, it could conceivably get through most of its life with a high and stable body temperature. A biophysicist called Jim Spotilla once calculated that a 2-ton lizard with

3. A term no longer in vogue in some taxonomic circles, but appropriately descriptive for our purposes.

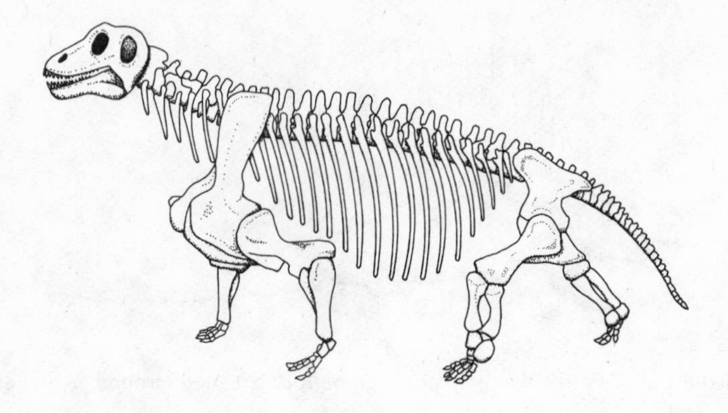

FIGURE 3.3a Skeleton of the enormous plant-eating Permian dinocephalian *Moschops*. Length around 5 meters (16.4 feet).

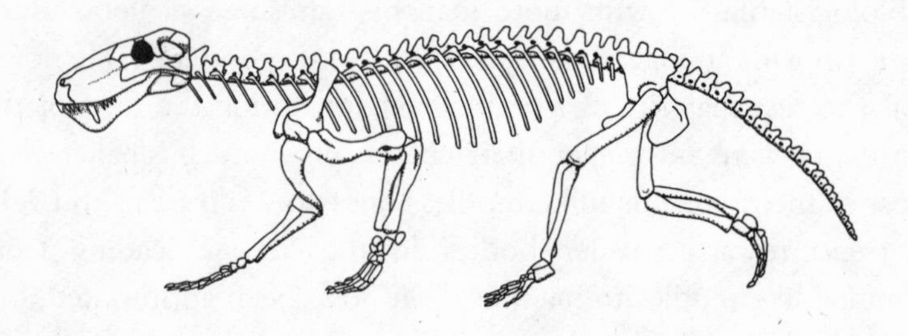

FIGURE 3.3b Skeleton of the carnivorous Permian gorgonopsian *Lycaenops*. Length around 1.4 meters (4.6 feet).

a typical reptilian metabolic rate living in southern Florida could keep its temperature between 30°C and 38°C for much of the year and never suffer from heatstroke or frostbite. It would be insulated against temperature change both by the climate and by its enormous thermal inertia. Hypothetical big cold-bloods such as Spotilla's are called "mass-homeotherms" in the jargon, where the term *homeotherm* indicates an animal capable of maintaining a relatively *stable* body temperature, regardless of what the actual set point is. The internal temperature of a 2-ton mass-homeotherm would not be as invariant as that of a mammal, but it would be considerably more stable than that of a 20-gram (.7-ounce) lizard subjected to the same environmental conditions. Add a fur coat, and a mass-homeotherm's thermal stability would be even greater. It is also likely that a lineage of furry mass-homeotherms would, over time, evolve an internal biochemistry that worked increasingly well within the narrow temperature range of their bodies. They would thus make a "commitment," in McNab's words, to a hoeothermic or temperature-stable way of life.

How does all this get us to the high metabolic rates of mammals? After all, the metabolism of our hypothetical mass-homeotherm is still firmly reptilian. Moreover, it seems unlikely that mass-homeotherms in warm and stable climates would ever evolve mammal-like rates of internal heat production: what would be the point if they had high and relatively stable body temperatures already? This is where the neat part of McNab's theory comes in. If, for whatever reason, natural selection began to favor the evolution of small-bodied species, a number of options would be available. Mammals could retain low metabolic rates and low levels of energy consumption as they shrank, but eventually they would have to abandon mass-homeothermy and return to controlling their body temperatures behaviorally. In other words, they would become like small lizards again. Alternatively, each successive decrease in body size could be accompanied by an increase in metabolic heat production to maintain the internal temperature to which their biochemical machinery had become accustomed. McNab contended that the scenario of increased metabolic heat production with decreasing body size could have been favored by natural selection. As the mass-homeothermic ancestors of the lineage had already molded their internal physiological processes to operate efficiently at a relatively stable temperature, perhaps the cost of returning to a more variable temperature was greater than the increased energy required to raise their resting metabolic rates.

Perhaps a commitment to temperature stability, once made, might be very difficult to undo.

In addition, competition with animals in the lineage leading to and including dinosaurs, for example, might have pushed those in the lineage leading to mammals into a nocturnal way of life. Night-time food resources were probably relatively underexploited at the time, and an internally warmed body would have allowed night hunting and foraging at will. Add enforced nocturnality to McNab's argument, and the evolutionary path from mass-homeothermy to full warm-bloodedness via a reduction in body size may well have been a constantly ascending one.

McNab's theory involves a few assumptions, but it is ingenious and at least consistent with the broad pattern of body-size variation known from the fossil record. It has received surprisingly little attention from scientists, however, partly because it is rather complicated but mostly because one year later, in 1979, a radically different theory was proposed that caught the imagination of most researchers in the field and seemed to offer a more general explanation for the evolution of warm-bloodedness in both mammals and birds.

Albert Bennett and John Ruben produced a theory about how warm-bloodedness evolved that has nothing to do with temperature at all, at least in its fundamental form. Their reasoning goes something like this. Man-made mechanical engines produce heat, and some produce enormous amounts of it, but no one would make the mistake of thinking that this was what they were designed for. Heat is a by-product—sometimes useful, sometimes a nuisance—of an engine's prime function, which is to turn the chemical energy contained within fuel into some sort of motion. This motion can be contained in one place to do work of various kinds (as in an electricity generator), or it can be used to impart motion to the engine itself (as in a car). The car analogy is of most use to us because animals carry their engines around with them.

In terms of origins, the important point about mammalian aspirated engines, say Bennett and Ruben, is not so much that they produce a lot of heat—although they do—as that they enable their owners to move around the environment at sustainable speeds that would exhaust a lizard in seconds. The potential advantages of better endurance—or a greater "aerobic capacity," in the jargon—are very obvious, which is one of the main reasons why Bennett and Ruben's theory is so attractive. Animals with high sustainable activity levels should be able to gather more food per minute, run farther in pursuit of prey or in avoidance of predators, outlast their less active peers in fights for food, space, burrows, or mates, prolong energetic sexual displays, patrol larger territories, and so on. In turn, it is not difficult to see how a high-capacity aspirated engine could, one way or another, lead to increased success in producing offspring, and thus how the requisite engine modifications could have been favored by natural selection. Most importantly of all, Bennett and Ruben's theory has no problem climbing the constantly ascending path: small increments in endurance would have required small increments in energy expenditure, but it is easy to see how the net benefits could have outweighed the costs throughout the entire journey from cold-bloodedness to warm-bloodedness. No valleys or peaks are involved, just a smooth uphill route.

Puzzle solved? Not quite. There is a hole in Bennett and Ruben's theory that no one has yet been able to plug. The problem lies not with the physiology of activity but with the physiology of inactivity. At peak output the aspirated engines of lizards and rats tick over at, say, 1000 rpm and 10,000 rpm respectively. However, when lizards and rats do nothing, their engines idle at 100 rpm and 1000 rpm respectively; much slower in both cases, but still with a tenfold difference. Ten times more food and oxygen is burned up inside a rat than inside a lizard for the purpose of doing nothing in particular. Surely the most cost-effective rat would be one with expanded top-end performance allowing it to sustain high levels of activity, combined with a fuel-efficient, lizardlike metabolic rate when at rest. Nothing in Bennett and Ruben's theory decrees that an animal with the capacity to be highly active has to have a high metabolic rate or a warm body when it is lounging around.

But, you might say, if natural selection progressively packed the cells of animals with greater numbers of mitochondria, or more powerful ones, in order to deliver better top-end performance, then surely this would result in a greater level of heat production at all times? Mammals have greater mitochondrial activity than lizards, so maximum metabolic rate and resting metabolic rate must be inextricably linked. Excess heat could just have been a coincidental by-product of selection for activity, and mammals simply trapped the waste heat under a fur coat in order to keep themselves warm.

If only life were that simple! The situation is far more curious and leaves Bennett and Ruben with some explaining to do. Mammals stay warm while at rest mostly because of the heat produced by metabolic reactions within our visceral organs, particularly the heart, kidneys, brain, liver, and intestines. These structures constitute only 7 percent of a human body by weight but account for 70 percent of our resting heat production. In contrast, the much higher power output we achieve while running is mostly a result of metabolic reactions that take place within our muscles. In other words, the aspirated engine of mammals is actually separated into two more or less distinct parts: the visceral subsystem that keeps us warm and the muscular subsystem that produces the top-end power we need for charging around. This uncoupling of the two systems means that there is no obvious reason why selection for top-end power should have had much effect on resting metabolic rate. A mammal with a lot of top-end power for sustained activity combined with a low resting metabolic rate for overall fuel economy still seems to be the most sensible design.

At the moment, we have no solution to this problem. Bennett and Ruben argue that maximum and resting metabolic rates *are* linked in some way so that selection for top-end power did result in an increase in resting metabolic rate, but the nature of this linkage remains unclear. What might it be? Ruben suggests one possibility. The major internal organs of mammals, particularly the kidneys, liver, heart, and brain, are much larger than those of reptiles of the same size. The rate at which metabolic reactions occur in the organs of rats relative to lizards is also higher by a factor of about four for the liver and kidneys and a factor of one and a half for the brain. The size and higher metabolic rates of these organs may be needed, Ruben argues, for the ancillary support functions of a highly active lifestyle, like coordination (brain), digestion (high activity requires more food), treatment and elimination of larger amounts of waste products (more food means more urine and feces), synthesis of larger volumes of chemical compounds, and so on. So the linkage between maximum and resting metabolic rates may be unavoidable. The total metabolic rate of the internal organs of mammals increased to support the rigors of an active lifestyle, and the heat produced as a by-product was harnessed for thermoregulation. Although speculative, the idea is at least entirely plausible.

We have explored two major ideas for the evolution of warm-bloodedness in mammals, one of which relies on the putative advantages of a warm body, and the other on the advantages of a higher aerobic capacity. Both ideas have attractive features and associated problems. Bennett and Ruben's theory is by far the most popular at the moment, but is there any empirical evidence to back it up? Hard evidence must necessarily come from the fossil record, but the hope of finding any has always been rather forlorn because the organs of greatest interest—hearts and lungs—do not fare well in the ground. Only very rarely do the soft parts of animals survive in rocks, and the hearts and lungs of mammal-like reptiles and early mammals are a complete mystery to us. Primary evidence about the transition from cold- to warm-bloodedness, therefore, if it can be accessed at all, will have to be extracted from a pile of old bones, and not a very big pile at that. Remarkably, paleontologists have managed to do just this. The evidence, and its crucial significance, revolves around what animals keep in their noses.

If you saw the film *Jurassic Park,* you may remember the scene where a velociraptor breathes on a window and fogs it up with condensation. The filmmakers could not have given a clearer indication that they considered their computer-generated stars to be warm-blooded (whether they were justified or not is another matter, and the subject of the next chapter). Warm-blooded animals are hot inside, breathe four or five times more often than cold-bloods of the same size, and expel large amounts of water vapor in the process. The only time this water catches our attention is when we are close to a pane of glass or venture out on a cold morning and see it condensing before our eyes. But whether it is seen or not, warm-blooded animals lose water from their respiratory systems at very high rates, and this is potentially a problem because animals weaken and die if they dehydrate, and warm-blooded animals breathe in and out so often that they are particularly vulnerable.

Mammals have evolved a rather neat partial solution to this problem, which we keep in our noses. Respiratory turbinates—RTs for short—are moist, often scroll-like bony or cartilaginous structures that project into our nasal cavities. As we breathe in, air flows over our warm RTs before it enters our lungs. As the air heats up, our RTs necessarily cool down. When we exhale, our RTs act like a pane of glass, cooling the outrushing air so that water condenses. The upshot is that RTs allow heat and, probably most importantly, water to be recovered. Ninety-nine percent of all living birds and mammals have RTs, but no cold-bloods do, a dichotomy that makes sense given our very high breath rates. So if we could look inside the noses of mammal-like reptiles and early mammals to see who did and did not have RTs, we might be well along the road to pinpointing not only the timing of the transition from cold- to warm-bloodedness, but also the kinds of animals involved, how big they were, where they lived, what they ate, and so on. Fossilized noses are, potentially, a paleophysiologist's time machine.

Unfortunately, although mammalian RTs are bony, they are also very delicate, so just like hearts and lungs they tend not to survive the process of fossilization. Unlike visceral organs, however, delicate bones sometimes attach to more robust ones and leave marks. Mammalian RTs attach to the bones of the nasal cavity and reveal their ghostly presence in fossilized skulls in the form of raised ridges. Willem Hillenius has recently made a survey of mammal-like reptile and early mammal skulls in search of RTs with some fascinating results. Figure 3.4 shows the temporal sequence of major groups of mammal-like reptiles and mammals. Pelycosaurs form the ancestral group arising in the Carboniferous, and these animals are thought to have given rise to therapsids in the Permian. Therapsids can be broken down into the five major subgroupings shown in the center of the diagram (dinocephalians to cynodonts). Mammals then arose from a cynodont ancestor sometime in the Triassic. The presence or absence of RT ridges in the skulls of these beasts, according to Hillenius, are: pelycosaurs: none found; dinocephalians: none found; dicynodonts: none found; gorgonopsians: none found; therocephalians: present; cynodonts: present; mammals: present.[4]

Although based on a relatively small number of suitable and

4. Here I follow Hillenius (1994). There is evidence that RTs may have been present in some gorgonopsians too (Kemp, 1969; see also Laurin, 1998), and work on this subject is being pursued by a number of people around the world at the present time. RT structures in earlier mammal-like reptiles cannot be discounted.

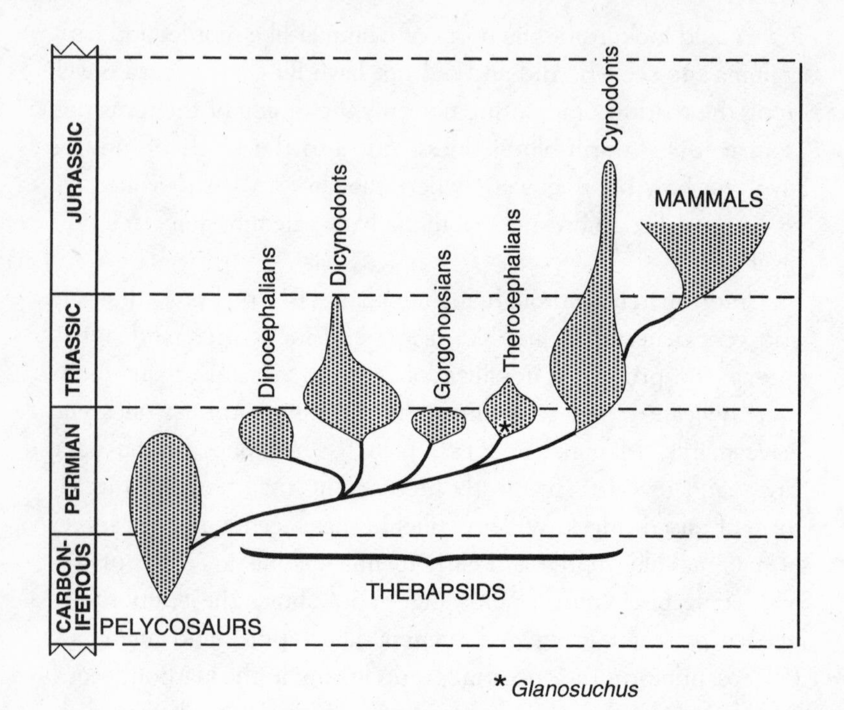

FIGURE 3.4 Geologic ranges and inferred evolutionary relationships of pelycosaurs, therapsids, and mammals. Redrawn from Hillenius (1994).

suitably prepared skulls—that is, not irretrievably squashed, confusingly distorted, or filled with rock—the pattern is highly suggestive. The therocephalian with a presumed RT ridge is *Glanosuchus,* a wolf-sized carnivore from the late Permian, 40 to 50 million years before the appearance of true mammals (Fig. 3.5). The RTs of *Glanosuchus* were probably small because its nasal cavity is too, and there are some other features in the roof of the mouth suggesting that it probably did not ventilate as fast as mammals. However, the nasal chambers expanded and the roof of the mouth became more mammal-like over the course of therocephalian evolution, which suggests that their breathing rates may

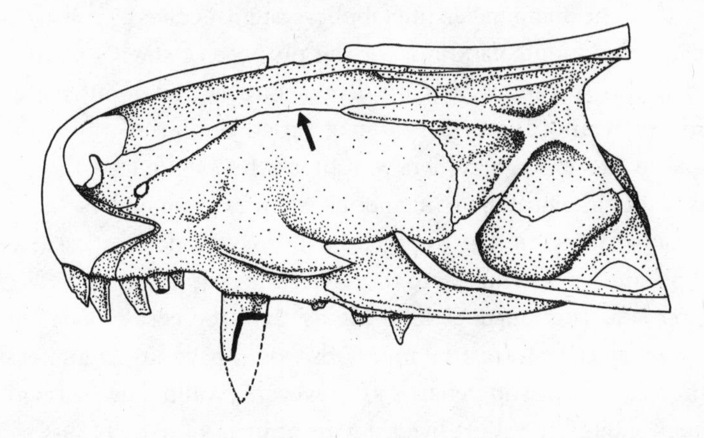

FIGURE 3.5 Section through the snout of *Glanosuchus*, a late-Permian wolf-like therocephalian. Arrowed is the ridge that may have supported respiratory turbinates. Depicted skull length 15 cm (5.9 inches). Redrawn from Hillenius (1994).

well have been increasing too. Lower- to middle-Triassic cynodonts and the earliest mammals have RT ridges comparable with those of living mammals, so they may have had broadly similar metabolic capabilities.

The RT ridges in *Glanosuchus* are particularly intriguing because this animal, like many therocephalians, was quite hefty and inhabited subtropical and tropical regions. Not only would *Glanosuchus* have been kept reasonably warm by the prevailing climate, it would also have benefited from the temperature stability that a large body inevitably provides. In other words, if incremental steps toward warm-bloodedness were under way in these large late-Permian animals, it seems unlikely that increased internal heat production for the purpose of keeping warm was the crucial factor. On the other hand, a large, environmentally warmed, mass-homeothermic predator might benefit enormously from incremental increases in top-end performance. So the hard evidence available at the moment tends to support Bennett and Ruben's idea that the mammalian metabolic system evolved, at least initially, for endurance and not for the purpose of staying warm.

There is still much that we do not understand about the evolution of warm-bloodedness, but a broad outline of the story is beginning to emerge. When vertebrates first colonized the land, they brought with them metabolic engines designed by natural selection for swimming (how could they have done otherwise?). But because of the influence of gravity the cost in energy of walking on land is about ten times higher than the cost of swimming in water, so early terrestrial tetrapods would have tired more easily than fish. Their superchargers, however, would have provided enough energy for short-lived, intense bursts of activity just as in living reptiles.

In the late Permian, natural selection began the process of souping up the metabolic engines of some therapsid reptiles. The best information we have suggests that the process may have begun in relatively large predatory therocephalian mammal-like reptiles in subtropical or tropical regions. The first steps were probably favored by natural selection because they increased the amount of energy available from oxygen-based metabolic reactions to fuel high levels of sustainable activity. Regulation of body temperature was probably not a factor initially. However, a more active lifestyle requiring more food, a higher rate of digestion, more rapid processing and elimination of waste products, and an increase in the rate at which molecules were manufactured within cells, may have led to a correlated increase in resting metabolic rate too. This elevation of maintenance metabolism came about partly through an increase in the absolute size of heat-producing organs like the heart, kidneys, liver, and brain, and partly because metabolic reactions proceeded faster within cells. With the addition of a fur coat and other control mechanisms, high and relatively stable internal temperatures were achieved at some point, but we don't know when. Evidence of gradually expanding noses in mammal-like reptiles hints that the transition to full warm-bloodedness started as much as 50 million years before the emergence of mammals and may have taken almost as long to complete.

So the evolution of warm-bloodedness, at least initially, had little to do with the benefits of warmth per se, and the process was well underway in reptiles long before the emergence of the warm-blooded animals with which we are so familiar today. Expansion of the standard reptilian aspirated engine to facilitate high levels of sustainable activity is clearly one of the most significant events in tetrapod history, but unfortunately this fact is too often taken to mean that the warm-blooded metabolic system represents some sort of fundamental advance over that of our cold-blooded ancestors. This view cannot be justified: warm-blooded and cold-blooded metabolic engines each have their associated advantages and disadvantages, and the ecological usefulness of each system depends on the size of the animal and the nature of the environment in which it lives. If the warm-blooded condition were inherently superior in the struggle for life, then the world would be crawling with mammal-like and birdlike creatures, whereas in truth it is quite literally crawling with lizards, frogs, snakes, spiders, and beetles. Even in the late Permian and on into the Mesozoic, when the world was largely unprepared for their emergence, mammals and their ancestors hardly seized the Earth with a burst of unalloyed competitive superiority. Quite the opposite, in fact. Pelycosaurs, dinocephalians, and gorgonopsians had all gone extinct by the end of the Permian. By the middle of the Triassic, *Glanosuchus* and its (warm-blooded?) therocephalian kin had all died out. Only two members of our once great lineage survived into the Jurassic, the prolific hairy cynodonts and the tiny early mammals. Cynodonts never recovered their earlier glory, dwindling in diversity through the Jurassic and becoming extinct before its end. And then there was one.

Oh, and a planetful of dinosaurs.

[4]

HOT AND COLD RUNNING DINOSAURS

When I was a child, dinosaurs were my introduction both to science and to awe. I still remember the books bought for me by my long-suffering parents, depicting reptilian giants like *Brachiosaurus* resting its chin on a pile of double-decker buses, fearsome *Tyrannosaurus* with teeth as long as a man's hand, and the mighty, spike-wielding *Stegosaurus* lumbering from tree to tree across the Mesozoic landscape guided by a brain the size of a walnut. Narnia and Wonderland were clearly worlds made up by adults for children, but the Mesozoic was a world of dragons and monsters beyond even a child's imagination.

I spent many hours living the lives of these ancient and fascinating creatures through books, and one in particular sticks firmly in my mind. The left side of each page showed a painting of a dinosaur in its natural habitat, while the right side gave facts and figures about the animal and some ideas about how it might have lived. At seven I was far more interested in the pictures than the words, so I tended to ignore everything on the right except for the animal's name. I remember this book so clearly because of one picture. It showed a *Brontosaurus* (Fig. 4.1) up to its shoulders in water in a small, vegetation-choked lake. Around the margins were marsh plants of various kinds and strange

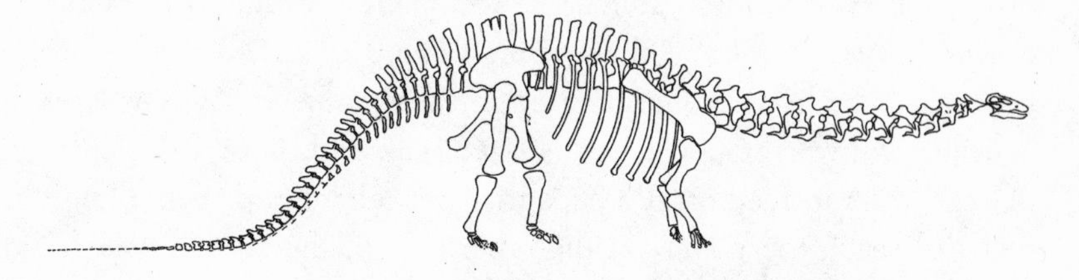

FIGURE 4.1 Skeleton of the late-Jurassic plant-eating dinosaur *Apatosaurus* (familiarly known as *Brontosaurus*). Length around 20 meters (66 feet).

overhanging trees dripping with water outlined against a bright blue sky. Dragonflies whirred across the surface. This was obviously a hot, humid, insect-ridden place like a jungle or a swamp, and the brontosaur's predicament was perfectly clear. An animal that big must have been very heavy—much heavier than an elephant, if the pile of double-decker buses was anything to go by—yet it had columnlike legs with feet almost as narrow as its knees. The poor creature must have stumbled into the fetid swamp by mistake. Once one leg had become mired, it would have been unable to lever itself out. Perhaps it had tried to swing its back end around to gain some leverage, only to become mired even deeper. Maybe it had tried to clamber to the far shore, or perhaps it had just given up the struggle and rolled over into deeper water. Whatever the grisly sequence of events, the unfortunate animal was now marooned in the middle of a lake either to sink inexorably into the stinking mud below or slowly starve to death.

It was a horrible picture to put in a children's book. For a long time I skipped that page altogether. I can't remember exactly how long it was before I got around to reading the text, but I'm ashamed to say it was probably years. When I did it was a revelation. *"Brontosaurus,"* said the right-hand side, "at 35 tons, one of the largest creatures ever to roam the Earth. Too heavy to walk on land, these enormous animals were aquatic, feasting on the soft, nutritious vegetation by the water's edge. . . ." Of course! The brontosaur hadn't become trapped in the swamp; this was where it lived. The water buoyed up its huge bulk and there was plenty of vegetation to feed on. Perhaps that was why it had such a long neck, to reach the plants on the other side of the lake as well as all those leaves on the overhanging trees. Mind you, it was an awfully small lake for such a huge creature; and there was no way of it getting out without sinking into the swampy vegetation; and there were no other brontosaurs for it to mate with; but perhaps the page simply hadn't been big enough for the artist to draw a larger lake and another brontosaur; and anyway, scientists know what they're talking about, don't they?

The irony, of course, is that my childish intuition turned out to be right, and for more or less the right reasons. Animals that habitually walk on yielding ground tend to have big ends to their legs. Coots and moorhens have enormous feet, camels have flaps of skin between their spreadable toes, and Canadians wear snowshoes when occasion demands. A 35-tonne (U.S. 40-ton) *Brontosaurus* had a total foot area of 1.2 square meters (1.4 square yards), so it must have supported around 29 tonnes on each square meter. An elephant holds up to 7.5 tonnes per square meter, cattle about 15 tonnes, tyrannosaurs 12 tonnes, and humans 2 tonnes. Fossil footprints show that some brontosaurs managed to walk on softish ground, just as elephants do today, but of all the animals that have ever lived, *Brontosaurus* would surely have been among the most tragic in a swamp.

Wallowing brontosaurs were ripe for reinterpretation in the 1970s, but they were not an isolated case. Most of the dinosaurs in my old picture books have been revamped in the last thirty years. *Brontosaurus* and *Brachiosaurus* (Fig. 4.2) have come out of the swamps to take their rightful positions as tree-browsing ani-

FIGURE 4.2 The late-Jurassic plant-eating dinosaur *Brachiosaurus*. Height around 16 meters (52 feet).

mals of dry land. Duck-billed dinosaurs (Fig. 4.3) have also emerged from the water to become grazers of low vegetation, the Cretaceous equivalent of buffalo. *Stegosaurus* (Fig. 4.4), that ponderous spike-tailed vegetarian with diamond-shaped plates running down its back, now rears up to browse from the tops of trees and pivots around on its enormous hind legs, waving its vicious spiked tail at the approach of any predator. *Tyrannosaurus* (Fig. 1.6), previously thought by many to be a slow-moving, cold-blooded behemoth, and perhaps no more than a scavenger of the

FIGURE 4.3 The late-Cretaceous plant-eating duck-bill *Edmontosaurus* (formerly known as *Anatosaurus*). Length around 13 meters (43 feet).

giant carcasses that must have littered the Mesozoic landscape, has woken from somnolent inertia to live up to its movie-star image as the most fearsome and terrible predator ever to walk the Earth. Three-horned *Triceratops* (Fig. 1.5), once the ponderous tank of the Cretaceous, now gallops away from its tormentors or, even more terrifying, toward them with its vicious cranial weapons lowered for battle. Dinosaurs have been reborn as beautifully adapted denizens of the Mesozoic environment, the hegemonic rulers of a fascinating Lost World.

The architect of this revolution in our understanding of dinosaurs was a brilliant, enigmatic, and rather eccentric paleontologist called Robert Bakker. In the 1960s Bakker got it into his head that the elders and betters of paleontology were wrong about dinosaurs, and he set out on a personal quest to prove it. He conducted his own research, collated and sifted data from other paleontologists, added ideas from biology and ecology, and eventually came up with a complete reinterpretation of the Mesozoic world. There is no doubt that Bakker is as much advocate as scientist, a duality that does not endear him to all of his colleagues, but as an accomplished and skillful practitioner of both arts he is simply impossible to ignore.

Of all Bakker's ideas about dinosaurs, one in particular has

FIGURE 4.4 The late-Jurassic plant-eating dinosaur *Stegosaurus*. Length around 8 meters (25 feet).

been the source of fierce controversy. In the early 1970s he made the bold claim that dinosaurs were warm-blooded, just like mammals and birds. A fully warm-blooded reptile, of course, is a contradiction in terms, so Bakker, true to his nature, went one step further and proposed (with a colleague, Peter Galton) a complete reclassification of the tetrapod family tree. Under their new scheme dinosaurs were grouped together in an official taxonomic category different from reptiles to the same degree that mammals are different from reptiles. (Previously dinosaurs were thought to have been members of two completely separate reptilian stocks.) Once Bakker and Galton had herded dinosaurs into their own taxonomic group, they went one step further and made them unextinct! Following on from the earlier work and ideas of the paleontologist John Ostrom, Bakker and Galton used the close anatomic similarities between birds and dinosaurs to justify the inclusion of the former as a subgroup of the latter. Birds are thus dinosaurs in the same sense that you and I are mammals, and while humans and birds roam the Earth, neither mammals nor dinosaurs can be extinct.

What prompted Bakker to make such bold claims about the metabolism of dinosaurs? After all, the only nonavian dinosaurs known to science have lain encased in rock for at least 65 million years. Bakker knew that he had to use multiple lines of reasoning to support his thesis and eventually he developed many, but the central idea was rooted in observations of the modern world. Recall our 100-km (62-mile) parade of living animals with a 5-tonne elephant at one end and a 0.000001-gram fairy-fly at the other. Starting from the elephant, the first 80 km (50 miles) of road are the exclusive domain of mammals. The first truly terrestrial reptile is a giant tortoise weighing in at only 250 kg (550 pounds). All the really big land animals on Earth are warm-blooded, and there are good reasons to believe that a fired-up metabolism is the underlying reason (see Chapter 6 for a full discussion). Large animals find it difficult to hide, which leaves them exposed to the elements, competitors, and predators. Small lizards can hide under rocks and in burrows to escape extremes of temperature and the attention of others, but rhino-sized animals simply cannot. The ability to stay at a constant temperature regardless of the weather, season, or climate, and to be highly active in competitive interactions with other animals must surely go a long way toward explaining the current hegemony of warm-blooded behemoths.

Mammals and dinosaurs appeared at about the same time in Earth's history, but dinosaurs came to dominate as large land animals for 140 million years, while mammals scurried around beneath their feet. The traditional view of dinosaurs as cold-blooded, therefore, suggests a time completely at odds with the world of today, a time when cold-blooded animals were large and warm-blooded animals small. Bakker concentrated his arguments around this apparent reversal of ecological principles in an attempt to shift the burden of proof onto his opponents. "The present is the key to the past" runs an old geological maxim that exhorts earth scientists to interpret the past in terms of the present unless there are good reasons for doing otherwise. Bakker pointed out that large animals today are warm-blooded and then asked his opponents to list the good reasons for thinking that the Mesozoic should have been any different. The tactic had considerable rhetorical weight because Bakker knew that dinosaurs had always been considered cold-blooded mostly because they had always been considered reptiles. To argue for cold-blooded dinosaurs from this position is reasonable only if dinosaurs were sufficiently like living reptiles to justify the classification in the first place, and this, as Bakker well knew, is a highly debatable point.

Even to the untutored eye there are obvious differences between dinosaurs and living reptiles. Reptiles are inveterate sprawlers with bodies held close to the ground, legs splayed out to the side, and heads more or less level with their bodies. Tortoises, lizards, and crocodilians all show this same low-slung style of carriage. (Crocodilians can elevate themselves into a semierect posture when walking, chameleons have rearranged their legs somewhat for a life spent walking along thin branches, and snakes and legless lizards have lost their legs altogether. These are all modifications to the basic reptilian plan rather than exceptions.) In contrast, dinosaurs stood and walked with their legs tucked in under their bodies like mammals, and many of them also held their heads high up in the air (Fig. 4.2). The significance of straight-up-and-down legs is still rather unclear,[1] but a fundamental insight into how

---

1. The paleontologist Gregory Paul, building on the pioneering ideas of Robert Bakker, contends that sprawled legs are well suited to the short stride-length and stable walking gait needed to move at slow speeds. If cold-blooded terrestrial tetrapods are to move continuously for long periods, they must do so relatively slowly so as not to exceed their aerobic capacity (see Chapter 3). The straight-up-and-down legs of mammals and dinosaurs are thought to permit longer strides, higher step frequencies, and thus speedier routine locomotion. If dinosaurs did routinely move at mammal-like speeds—and there is evidence from fossil footprints that they did—then this would be a good indication that they had aerobic capacities well above those of any living reptile.

these animals must have worked internally comes from the fact that many dinosaurs managed to elevate their heads well above the level of their bodies.

Imagine a garden hose with one end attached to a tap and the other end free. With the hose held level to the ground and the tap turned on, the water will gush freely over the lawn. Hold the free end up in the air, however, and the force of the emerging water will lessen. Hoist the free end on a long stick, and eventually the weight of water in the hose will balance the pressure from the tap and the flow will stop altogether. Now imagine that the hose is an artery and the tap an animal's heart. For a low-slung crocodile, very little force is needed to get blood from heart to head. Reptiles pump blood out of their hearts at a pressure of 35 to 75 mm of mercury (mm/Hg) above the pressure of their tissues, a differential that is necessary mainly to drive blood through narrow capillaries in the peripheral regions of their bodies. Humans are like crocodiles rotated through 90 degrees, so we need more powerful hearts capable of pumping blood uphill; hence the characteristic human blood pressure of 100 to 150 mm/Hg. To get blood to a giraffe's head requires a staggering pressure of 300 mm/Hg. The only logical extension of this line of reasoning is that *Brachiosaurus* (Fig. 4.2), with a head 13 meters (43 feet) off the ground and 8 meters (26 feet) above the rest of its body, must have had a heart quite unlike that of any modern reptile and probably among the most powerful in animal history. One paleontologist has even speculated that a heart capable of getting blood around a *Brachiosaurus* must have weighed around 1.5 tonnes!

And we can extend this line of reasoning a little further. The heart of an animal has to pump blood not only to its head and feet but also to its lungs, where oxygen is taken up and carbon dioxide expelled. Lungs are delicate organs with thin-walled capillaries running very close to the surface to aid the inward and outward exchange of gases. If a giraffe pumped blood to its lungs at a pressure of 300 mm/Hg, the blood vessels at the lung surface would rupture in spectacular and terminal fashion. Mammals and birds have overcome this problem by physically separating their hearts into right and left sides, each being responsible for propelling blood to a different part of the body. Blood enters the right side and is pumped out under low pressure to the lungs. Oxygen and carbon dioxide are exchanged, and the blood then returns and drains into the left side. The pumping chamber on the left has much thicker walls than on the right, so when the heart contracts, blood is forced out at high pressure to the head, arms, legs, and internal organs. The blood then returns to the right side of the heart and the cycle starts over. The pumping chambers of a reptilian heart, in contrast, are not completely cut off from each other (the hearts of crocodilians and lizards, for example, differ in the degree of physical separation between chambers, but both maintain a connection). This system gives reptiles a range of options for controlling their circulation. While lizards are basking in the sun, for instance, they can shunt blood away from the lung circuit and into the body circuit to warm it faster. Similarly, blood can be shunted preferentially to the body while diving after the store of oxygen in the animal's lungs runs out. After diving, blood can be shunted the other way to do repeat circuits of the lungs to increase the rate of reoxygenation. The reptilian heart may not be very powerful, but it is a marvelously versatile organ.

What sort of heart did *Brachiosaurus* have? It must have been able to deliver an almighty push to get blood up to the animal's head, but it must also have been able to separate the lung and body circulation into low- and high-pressure circuits. No one knows exactly how their hearts were constructed, but power combined with the capacity to produce an enormous pressure differential must have been the most important design criteria. Brachiosaur hearts were probably like those of mammals and birds. And then some.

What about the respiratory system of dinosaurs? Bakker suggested that this too might have had more in common with birds than with reptiles. Bird lungs consist of a series of parallel tubes through which air flows continuously in the same direction whether the bird is breathing in or out. Air flows not only through the lungs but also through a complex series of air sacs that occupy large areas of the body cavity and even extend into the skeleton. The fact that skeletal air spaces are intimately connected to the respiratory system was demonstrated in 1758 in an unconscionably gruesome experiment by John Hunter, who blocked the windpipes of chickens and hawks and then cut through their wings. He found that the birds could still fill their lungs with air by breathing through their severed wing bones. Bakker noted that the vertebrae of brontosaurs and some carnivorous dinosaurs also contain cavities like those of birds and concluded that dinosaurs probably had similar respiratory systems.

And the link between dinosaurs and birds extends well beyond some similarities in bone structure. Debate has raged for many years—ultimately since Thomas Huxley in the late nineteenth century—about whether birds evolved directly from dinosaurs or whether both birds and dinosaurs arose from a much older ancestor that could not be classified as either. The dinosaur-bird advocates have had the upper hand for some time now, and recent fossil discoveries in China have strengthened their hand even further. In 1998 two scientific papers were published describing small carnivorous dinosaurs sporting what appear to be featherlike structures. The first was named *Sinosauropteryx prima* ("first Chinese dragon-feather," Fig. 4.5). *Sinosauropteryx* had downlike structures on its body, but it was clearly not a bird and equally clearly could not fly. Hard on the heels of this publication came descriptions of

FIGURE 4.5 Might the small theropod dinosaur *Sinosauropteryx prima* have looked something like this in life? Fossils recently excavated in China suggest that this dinosaur and others sported downlike and/or featherlike structures. Drawing based on the photographed model of *Sinosauropteryx* in Ackerman (1998).

Protarchaeopteryx and *Caudipteryx*, both with downy coverings like *Sinosauropteryx* but also with vaned feathers on their tails and arms. Although *Protarchaeopteryx* and *Caudipteryx* had feathers like modern birds, they were probably ground-dwelling running animals without the power of flight.

What use would nonflying dinosaurs have for feathers? Insulation is one possibility, but they may have been used primarily for camouflage or display.[2] If the feathers were for insulation, this would be good evidence that these dinosaurs at least were warm-blooded (the only warm-blooded animals alive today without ex-

2. There is a simpler interpretation of the feathers on *Protarchaeopteryx* and *Caudipteryx,* however, supported by a number of paleontologists. These dinosaurs may, in fact, have been secondarily flightless (i.e., their ancestors could fly, but at some point the power of flight was abandoned). All living flightless birds have either feathers or the vestiges of them, so feathers on secondarily flightless dinosaurs would be unsurprising.

FIGURE 4.6 *Hypsilophodont,* one of a group of ostrichlike plant-eaters extant from the mid-Jurassic to the late Cretaceous. From 90 cm to 2.5 meters (2.9 to 8.2 feet) in length.

tensive coverings of fur or feathers are either big, like elephants, fat, like pigs, aquatic, like whales, or naked mole-rats). Even if the feathers did not serve as insulation, the dividing line between dinosaurs and birds these days is becoming increasingly blurred. As dinosaurs appear to be so closely related to animals that we know to be warm-blooded, the possibility that some dinosaurs might have possessed a birdlike metabolism does not seem a particularly wild conjecture.

These anatomic features and evolutionary relationships are good reasons for setting dinosaurs apart from living reptiles. However, none of the arguments presented so far demonstrate with any force that they were warm-blooded. Dinosaurs probably had the internal equipment *capable* of supporting a high-powered metabolic regime, but this does not necessarily mean they were like mammals and birds. Other lines of evidence were required, and they were quick to come. Bakker noted that large dinosaurs, like brontosaurs, duck-bills, stegosaurs, and horned dinosaurs, had relatively long legs with joints similar to those of rhinos and elephants, and he inferred from this that walking and running speeds were probably comparable. Hypsilophodont dinosaurs (Fig. 4.6) resembled ostriches, with very long, graceful limbs clearly built for running at speed. Bakker calculated that the maximum sustainable speed for a cold-blooded 100-kg (220-pound) hypsilophodont would have been 3 km (1.9 miles) per hour. Why would a dinosaur be built like an ostrich if it could only lever itself across the landscape at 3 km per hour? Dinosaurs, Bakker argued, were built for sustained speed, and this requires the sort of high-powered metabolism possessed by mammals and birds.

Other evidence turned up in some unlikely places. In the 1960s oil geologists working in north Alaska began to unearth ceratopsians (related to the three-horned *Triceratops*), carnivorous dinosaurs, duck-billed dinosaurs, and hypsilophodonts. Continental landmasses have occupied different places on the globe at different times in the past, but this part of Alaska was well within the Arctic Circle even during the Cretaceous. Since these early discoveries dinosaur remains have turned up in other parts of Alaska, Bylot Island north of Greenland, Svalbard, and in Cretaceous sediments in northern Eurasia. Most dramatically of all, dinosaurs have been found in Jurassic and Cretaceous rocks from Antarctica and Victoria in south Australia, areas located well within the Antarctic Circle at the time the animals died. Polar dinosaurs? Undoubtedly. But *cold-blooded* polar dinosaurs? Without a high level of internal heat production, how could they possibly have survived? About 20°C (68°F) is the minimum temperature at which food can be digested. Bakker and his supporters saw this biogeographic evidence as another nail in the coffin for the idea that dinosaurs had reptile-like metabolic engines.

Further evidence came from microscopic analysis of dinosaur bones, which seemed to reveal an internal structure most closely resembling the bones of mammals. Bone in reptiles is usually compact, but the bones of mammals and birds typically have a fibrous, woven appearance with dense networks of channels where bone has been reabsorbed, laid down, and reabsorbed again. This sort of structure is broadly characteristic of animals that grow quickly, and high rates of growth tend to correlate with high rates of metabolism. A dog grows from 2 kg to 60 kg (4 to 132 pounds) in a year. Lions take about two years to reach 100 kg (220 pounds). Ostriches grow even faster, from egg to 70-kg (154-pound) adult in only nine months. Mississippi alligators, however, take ten to twenty years to reach 100 kg in the wild, and reticulated pythons may take a similar time to reach 55 kg (121 pounds) even in a zoo. Dinosaurs had bone with fibrous texture, so, ran the argument, they were probably fast growers. Fast-growing extant animals are characteristically warm-blooded, so dinosaurs probably were too.

A particularly intriguing line of evidence came from chemical analyses of dinosaur bones. Reese Barrick and colleagues measured the ratio of oxygen isotopes in the bones of a lizard and four dinosaurs from late Cretaceous rocks in Montana. Isotopes of oxygen differ in weight because they contain different numbers of neutrons, and the extent to which heavy oxygen and light oxygen atoms are incorporated into bone depends on temperature. Because the chemical composition of dinosaur body fluids is not known, Barrick could not estimate the internal temperatures of his specimens, but he could infer how their body temperatures *varied* from the variation in isotope ratios. A few years earlier another group of researchers had modeled the effect of climatic variation on a hypothetical cold-blooded 5-tonne duck-bill from the same rock formation, thought to have formed under a subtropical or warm-seasonal climate, and predicted that the animal's core body temperature would have fluctuated between 34°C (93°F) in July and 12°C (54°F) in December. A warm-blooded dinosaur, on the other hand, would have been able to keep its temperature constant all year round. If cold-blooded duck-bill deposited bone continuously, Barrick reasoned, then the isotope ratios in its bones should reflect its wide range of body temperatures, but if it had been warm-blooded, the isotope ratios should be more stable.

Analysis of a meter-long (39-inch) Cretaceous lizard suggested that its body temperature fluctuated by 10 to 15°C (18 to 27°F), about right for cold-blooded animals depositing bone in a seasonal climate. But the results for the dinosaurs were different. The variability in isotope ratios for a 1-meter juvenile hypsilophodont was half that of the lizard, and for the other three specimens—ranging in size from a 2-meter (6.6-foot) ceratopsian to an 8-meter (26-foot) duck-bill—about half as much again. The authors concluded that all the dinosaurs had metabolic rates above those of living and extinct reptiles and that they regulated their internal temperatures within quite narrow limits, more like mammals than like reptiles.

Bakker himself came up with a particularly ingenious line of reasoning based on the ecological consequences of differences in metabolic rate and the relationship between predators and their prey. Imagine a herd of 700 antelopes, each weighing 50 kg (110 pounds). All this meat-on-the-hoof might support 300 predatory lizards also weighing 50 kg each. The predator–prey ratio of this system, therefore, would be 0.3 or 30 percent by weight (300/1000; the sum of predator and prey is used in the denominator because the predators can be scavenged when they die). However, because of the high metabolic rates of mammals and their consequent high food demands, the same herd of antelopes would probably support only around 40 similar-sized predatory cats, which yields a predator–prey ratio of 0.05 or 5 percent. So, Bakker reasoned, if we could weigh all the individuals in a predator–prey system, the proportion formed by the meat-eaters should be higher if they are cold-blooded than if they are warm-blooded. Of course, weighing a predator–prey system is not possible in practice, and even making reasonable estimates is fraught with difficulties, but the inescapable logic of the link between metabolic rate and predator abundance was just too much for an enthusiast like Bakker.

The results of his extensive studies on this subject are very interesting. Ratios[3] for living cold-blooded species (twenty-nine studies of invertebrates and seven of vertebrates) were always above 10 percent and sometimes as high as 60 percent, while those for warm-blooded species (eighteen studies of mammals) were always below 10 percent, just as Bakker predicted. In the early Permian period (Fig. 3.1) the land was dominated by cold-blooded reptiles and amphibians, the most famous being the fin-backed pelycosaur *Dimetrodon* (Fig. 3.2). Predator–prey ratios in early Permian rocks were high, usually above 40 percent. Bakker then went on to study Cenozoic (less than 65 million years old) fossil mammals from America and found predator–prey ratios all below 5 percent. So far, so good. But what about dinosaurs? Bakker totted up animals in rock formations from China, Europe, Africa, and America and found that predatory dinosaurs were conspicuously rare. The ratios? Between 2 percent and 6 percent, indistinguishable from those of living and fossil mammals. Bakker declared that dinosaurs were similar to mammals in their predator–prey ratios, so they must have been similar in their physiology as well.

All in all, Bakker and his supporters put together a cogent, elegant, and empirically supportable case. The skeletal anatomy of dinosaurs suggests that they were robust, active animals. In posture, gait, and their hearts, dinosaurs were completely unlike living reptiles but similar to large living mammals. Dinosaurs had bones that seem to indicate fast growth and stable body temperatures.

---

3. For living animals, Bakker used productivity efficiencies, defined as the energy contained in all the tissue produced by a population by growth and reproduction divided by the total energy in the food consumed by the population minus that lost through defecation and regurgitation. In theory, Bakker argued, productivity efficiencies (expressed as a percentage) and predator–prey ratios from fossil deposits (also expressed as a percentage) should be similar.

Fossils have been unearthed at polar latitudes where living cold-blooded animals would perish, and predator–prey ratios among dinosaurs were the same as among both living and extinct mammals. Surely Bakker is right: dinosaurs *must* have been warm-blooded.

Convinced? As an undergraduate paleontologist I was. But professional paleontologists are a suspicious bunch. Paleophysiological inferences are based on layer upon layer of circumstantial evidence, and every layer has to be judged carefully on its own merits. Theories about the workings of extinct creatures are invariably complex and always assumed guilty until proven innocent. It takes more than cogency, internal consistency, and empirical weight to change the collective mind of a community of professional skeptics.

Critics were quick to point out, for example, that the Mesozoic climate was on average considerably milder than today, with tropical and subtropical conditions extending much closer to the poles, so there may have been no real advantage to being warm-blooded. In fact, cold-bloodedness might have been better. Their bulk[4] and the prevailing climate would have kept dinosaurs warm, so they would not have had to expend large amounts of energy in thermoregulation. To counter this argument, Bakker pointed out that although a mass-homeothermic dinosaur in a warm climate would have enjoyed greater thermal stability than a 20-gram (.7-ounce) lizard, it would not be able to maintain a high and constant body temperature at all times like a mammal. A drop in body temperature from 38°C to 30°C would lead to a reduction in the rate of physiological processes somewhere in the region of 40 percent.

4. Most of the dinosaurs we know about are large, although some, like *Compsognathus,* were not much bigger than a chicken. There were probably many more small species, but their delicate skeletons were much less likely to be preserved than those of giants like *Brontosaurus.*

FIGURE 4.7 The fearsome Komodo dragon. Males may occasionally exceed 3 meters (9.8 feet) in length.

Mass-homeotherms, Bakker argued, would surely lose out in the long run to warm-blooded animals in contests like digesting, growing, running, mating, and fighting. But then again, dinosaurs never *had* to compete with large mammals in the Mesozoic. For whatever reason, dinosaurs managed to evolve into large animals while mammals remained small. It is one thing to expect large mammals to outcompete mass-homeotherms in head-to-head competition, but quite another to expect mouse-sized mammalian insectivores to evolve into elephant-sized herbivores and carnivores when around every corner stands a mass-homeotherm with sharp teeth, a mean disposition, and a surprising turn of speed. Bakker (1986) himself concedes a similar point when he tries to explain how cold-blooded Komodo dragons (called *ora* by locals; Fig. 4.7) managed to evolve into such fearsome large predators: "The dragon succeeds only where it is free from interference from large mammal predators. . . . On Komodo Island not one large mammal predator has ever existed in the wild. (Natives keep dogs on Komodo, but these canines are a wretchedly scrawny lot, hardly a threat to the ora)."

If dogs pose no threat now that Komodo dragons are so big and mean, how much of a threat would a mouse-sized nocturnal insectivore be to a mass-homeothermic dinosaur? On the other hand, could mass-homeothermic dinosaurs really have dominated so completely for 140 million years that mammals were denied *any* chance to evolve into large animals? Bakker says no. Others say yes. It seems to be a matter of faith.

Climatic differences could also explain the existence of polar dinosaurs, claimed the critics. There is little evidence of ice and snow at the poles at the time dinosaurs were there, so perhaps the climate was within the tolerance limits of cold-blooded animals. Certainly the cool, dark winters would have been a problem, but many animals migrate to escape the worst excesses of the climate, and dinosaurs may have done so too. And if dinosaurs migrated, then they may have been able to keep their body temperatures more or less constant all year round, which might explain Barrick's isotope results. Dinosaurs may have been like caribou—albeit rather slow ones—migrating north to feed on the summer flush of vegetation, then moving south with the onset of winter.

Bakker's opponents also pointed out that low activity levels and cold-bloodedness do not necessarily have to be causally linked, although among living vertebrates there is certainly a strong correlation. Some small lizards under laboratory conditions have been found to be as active as small mammals. Fish can be highly active, as can mosquitoes and houseflies. Anyone who has witnessed the strike of a rattlesnake or faced down a crocodile on land will know that cold-blooded animals can move extremely fast if they want to. Certainly living reptiles cannot sustain high levels of activity aerobically for as long as mammals, but how long does it take to escape into the bushes? And how far would a cold-blooded herbivore have to run if its predator was also cold-blooded? Limitations on stamina would have applied to predators and prey alike. Dinosaur chases may have been relatively short affairs, but they may also have been intense, physically demanding, and no less crucial for the hereditary line of predator or prey. That some dinosaurs were built for speed is surely true, but Bakker's contention that they were built for *sustained* speed shifts the debate into a rather gray area.

Bone-texture studies were the next to come under fire. Tomasz Owerkowicz compared the bones of active and sedentary lizards and found that the well-exercised group developed "fast growth" bone, suggesting that this type of texture might be characteristic of an active lifestyle rather than of warm-bloodedness per se. Bone with fibrous texture is also absent from many small birds and mammals, which suggests that size or lifespan may be important. The correlation between bone texture and warm-bloodedness is thus far from absolute, and some of the evidence suggests that the link may be related to growth rate, metabolism, body size, activity levels, bone strength, or some combination of factors.

And bone-texture studies have also been used to support the view that dinosaurs were cold-blooded. A number of dinosaur fossils show bone deposited in zones separated by darker rings called lines of arrested growth. These may indicate seasonal variations in growth, a pattern common in living cold-bloods because they grow slowly, if at all, in winter. As warm-bloods remain at the same temperature all year round regardless of the climate, their rates of growth tend to be more constant. Again, however, there are exceptions. Some warm-blooded animals produce lines of arrested growth during periods of malnutrition and after they have reached adult size, while some cold-blooded animals seem to lack them altogether. We simply do not know enough about what causes differences in bone texture to be sure what the differences and correlations mean.

What about the dinosaur-bird connection? The majority of paleontologists suspect that carnivorous dinosaurs were the direct ancestors of birds. If this is true, then warm-bloodedness must have arisen at some point in the lineage connecting dinosaur ancestors, dinosaurs, and birds, but the big question is, where? Even if the birdlike dinosaurs *Caudipteryx* and *Protarchaeopteryx* were warm-blooded, it tells us nothing about *Triceratops, Brontosaurus, Stegosaurus,* and other distant twigs on the dinosaur evolutionary tree. Demonstrating that some dinosaurs were warm-blooded is a long way from demonstrating that all, or most, or any but a specialized handful were.

Predator–prey ratios? While paleontologists praised Bakker's insight and exhaustive studies, they had little difficulty raising objections. Grappling with an incomplete and fickle fossil record is what paleontologists do for a living, and the difficulties involved in any sort of paleoecological reconstruction are widely appreciated. The number of animals that turn up as fossils is minuscule compared with the number that must have lived. How can we be sure that those animals that did get preserved died *in situ*? What if a load of bones was washed in from another area when a river broke its banks? What about the scattering effect of scavengers? What if predatory dinosaurs lived much longer lives than herbivorous ones and so contributed their bones to the pile less often? Bones of theropods (the predators) also tend to be thinner and more delicate than those of the large herbivores on which they preyed, so we would not expect them to preserve as well or as frequently. Are patterns of mortality among large land animals predictable enough in time and space for predator–prey ratios to mean anything at all? There are just too many unknowns, claimed the critics, for them to be conclusive evidence of anything.

By this point you may be thinking that the advocates of cold-blooded dinosaurs are little more than spoilsports. It is certainly true that Bakker and his supporters were the proactive group for a long time and that the cold-blooded camp sat back and picked holes in their arguments. Criticism is the backbone of science so there is nothing wrong with this tactic, but positive evidence for the claim that dinosaurs were cold-blooded would be much more convincing. The evidence has been a long time coming but it has now arrived, and it revolves, once again, around those structures called respiratory turbinates that some animals keep in their noses.

John Ruben and colleagues put the skulls of two carnivorous dinosaurs and a duck-bill into a medical CAT scanner, which allowed them to generate a 3-D computer image of the skulls' internal structure. They found no trace of RTs. This is perhaps not surprising, because RTs are delicate structures that could easily disappear during fossilization. To accommodate RTs, however, the nasal passages of mammals and birds are, on average, about four times larger in cross section than those of living reptiles. The CAT scan allowed Ruben to measure the cross sectional area of the nasal passages and on this point the results were conclusive: dino noses fall squarely within the size range for lizards and crocodiles and well below the range for mammals and birds, which suggests, claim the authors, that dinosaurs were cold-blooded.

A year later Ruben and colleagues published another paper which seemed to slam the door shut on the idea of warm-blooded dinosaurs. We have already met *Sinosauropteryx* (Fig. 4.5), the enigmatic carnivorous dinosaur from China that seems closely related to birds, but Ruben thought he saw something in this fossil that suggested a closer link with crocodiles. Both mammals and crocodiles have a sheetlike diaphragm separating the heart and lungs from the liver and other internal organs. The diaphragm is constructed differently in crocodiles and mammals, but it has the same function: backward movement of the crocodilian diaphragm (downward in upright humans) causes the frontal cavity to expand, which, in turn, sucks air into the lungs.[5] As mentioned earlier,

5. Mammal lungs consist of millions of tiny sacs (alveoli), which are richly supplied with blood vessels. When the diaphragm of a mammal contracts and the lung cavity expands, all the alveoli expand like tiny bellows. Breathing out is accomplished mainly by elastic rebound of these structures. This type of lung functioning, combined with a rich blood supply and a very thin blood-air barrier, allows mammals to garner enough oxygen to sustain high levels of activity. A reptile lung, in contrast, is like a single

birds have a different type of respiratory system and, crucially, they do not have a diaphragm at all. When Ruben saw the *Sinosauropteryx* fossil from China, one feature immediately grabbed his attention: this birdlike dinosaur had a body cavity that seemed to be separated into forward and backward halves, just like a crocodile. If the lungs of *Sinosauropteryx* were simple diaphragm-operated bellows like those of crocodiles, then the simplest conclusion is that these animals were probably incapable of the sort of exercise-related rates of gas exchange typical of birds. In other words, if *Sinosauropteryx* had an internal organization like that of a crocodile, then the most straightforward interpretation is that it probably had a crocodile-like metabolism as well.

How did the advocates of warm-blooded dinosaurs respond to this new evidence? Well, they now had the long-awaited chance to pick holes in the positive evidence presented by the opposition, and they had little trouble doing so. For a start, just because a fossil does not show evidence of respiratory turbinates, it does not follow that the animal didn't have them. Very little organic material of any kind gets preserved in rocks. The evolutionary history of mammals, for example, has been pieced together largely on the evidence of teeth, which are the only bits of a mammal resistant enough to stand a fair-to-good chance of being fossilized. If skulls and thigh bones rarely survive the process of fossilization, what chance have wispy slivers of bone? And the RTs of birds, and perhaps therefore dinosaurs, are composed mainly of cartilage, which would have even less chance of being preserved. Moreover,

---

enlarged mammalian alveolus with ingrowths (septae) that separate the lung into compartments (a cross section looks a bit like a cartwheel). Gas exchange takes place mainly on the septae. Reptile septae are less densely supplied with blood vessels than the alveoli of mammals. Large areas are hardly vascularized at all and just assist with ventilation of the lung as a whole. Although mammal and reptile lungs are essentially bellows that suck air in and then reverse the flow to push it out again, differences in internal structure and function mean that the mammalian lung is much more efficient at gas exchange.

whales and elephants seem to get by with no RTs without any problems. And kiwis have exceptionally narrow nasal passages—much narrower than Ruben's dinosaurs—yet they *do* have RTs. Perhaps dinosaurs just had other ways of limiting water loss. Perhaps they drank more. We know that dinosaurs employed the relatively water-efficient uric acid-based system of excretion like most living reptiles rather than the more wasteful urea-based system used by mammals. If dinosaurs had other means of regulating water loss, then they may have needed neither RTs nor expanded nasal passages. RTs may also act as brain coolers in big-brained animals like mammals and birds, but most dinosaurs had relatively small brains that may not have needed cooling. Did RTs evolve as water regulators, brain coolers, or both? These objections may seem vague and picky, but it is the warm-blooded advocates' turn to be picky.

As we saw in the last chapter, some mammal-like reptiles had RTs—or at least the sites of attachment are visible in some skulls—which suggests that some if not all of these animals were warm-blooded. If the supporters of dinosaurian cold-bloodedness are right, therefore, the historical sequence of dominant large land animals on Earth would start with warm-blooded mammal-like reptiles, move on to cold-blooded mass-homeothermic dinosaurs, and then to warm-blooded mammals. Is such a sequence likely? The advocates of warm-bloodedness say no: if cold-blooded-mass-homeotherms could match warm-blooded animals in head-to-head competition, then where are all the 2-tonne reptiles in our present world? The tropics at least should be crawling with them. Others say that this argument is fundamentally flawed because, as the warm-blooded advocates are fond of pointing out for their own purposes, living reptiles are not good models for dinosaurs. And so the argument goes round and round.

What about the Chinese *Sinosauropteryx* and its supposedly crocodilian diaphragm? Surely this crucial specimen should have the warm-blooded advocates mumbling sheepishly into their coffee. Not a bit of it. *Sinosauropteryx*'s piston ventilation system *as reconstructed by Ruben* does resemble that of a crocodile, but it also resembles that of a mammal. No other living reptile has a crocodile-like piston; lizards, for example, manage to aerate themselves quite adequately just with expandable rib cages. It is widely accepted that the diaphragm-assisted ventilation system of mammals facilitates the high rates of gas exchange needed to fuel a warm-blooded metabolism. *Sinosauropteryx's* piston may have performed the same job. What we really need to see is *Sinosauropteryx's* lungs. Unfortunately, the probability of finding a dinosaur with its exquisitely delicate lung tissue preserved is not significantly different from zero. Without information on the lungs, the function of *Sinosauropteryx*'s pistonlike ventilation system is impossible to gauge. What is more, where some paleontologists see a line on the *Sinosauropteryx* specimen separating forward and backward body cavities, others see a breakage in the rock caused by the farmer who disinterred it with rather more enthusiasm than paleontological know-how. A breakage, moreover, that the enterprising fellow then mended with colored cement and glue. Who has been fooled? No one yet knows for sure. Fortunately, bird-dinosaurs are turning up in China with unprecedented frequency, so we may have the answer soon. Some of us can't wait.

What is the upshot of thirty years of study and fierce debate? Only that metabolic paleontology continues to fascinate and stretch some of the best and most ingenious minds in science. Ecology, biology, paleontology, geology, biogeography, climatology, isotope geochemistry—what other discipline can boast of practitioners able to draw on such a breadth of knowledge and expertise? The debate has been long, tortuous, and sometimes acrimonious, but to me it has been a triumph of the scientific spirit. There is no answer to the original question as yet, and perhaps there never will be, but the debate has stimulated scientists in many different disciplines to think again, and more deeply than ever before, about some of the most fundamental issues in biology. When pushed by inquisitive paleontologists, physiologists and biochemists have had to admit that they don't really understand how warm-bloodedness works. Histologists don't really understand why bones look the way they do. Evolutionary biologists aren't sure how or why warm-bloodedness evolved. Ecologists and biogeographers have had to think again about how metabolism affects population size and growth, environmental tolerances, geographic distributions, habitat relationships, competitive interactions, predator–prey systems, community organization, responses to environmental change, and much more besides. Most scientists specialize in the present, but by resurrecting strange and magnificent beasts from the Earth's distant past, paleontologists have forced scientists to think beyond the actual to the possible, to set the tiny, unrepresentative time-slice of the present within the grand sweep of history. The discipline can be proud of these achievements, even if an answer to the original question remains elusive.

What sort of evidence might clinch the issue? More predator–prey studies and a better understanding of bone texture would be useful, but given the uncertainties, I cannot see either approach yielding definitive answers. If RTs turned up in dinosaurs the case for warm-bloodedness would be strengthened considerably, but repeated demonstrations of their absence will tell us little more than we already know, which is that we don't understand how dinosaurs balanced their water budgets. Well-controlled isotope studies of reptile, mammal, and dinosaur bones from the same deposits would be highly informative. If oxygen–isotope ratios of dinosaur bones consistently matched those of reptiles and not mammals, the warm-blooded scenario would be significantly weakened. Incontrovertible evidence of reptile-like lungs in *Sinosaurop-*

teryx and its kin would have the same effect. It is a fascinating situation and beautifully poised for new research and fossil discoveries.

The warm-blooded/cold-blooded dichotomy has dominated the debate about dinosaur physiology, but perhaps the most exciting possibility, and one that has become increasingly popular among paleontologists in the last decade, is that dinosaurs were neither souped-up crocodiles, nor monstrous birds, nor mammal-analogues, but something unique and unlike anything currently alive on Earth. Perhaps they had metabolic rates between those of crocodiles and mammals. Perhaps they had variable rates or different metabolic characteristics in different lineages. What would an active 10-tonne cold-blooded predator with a high-pressure heart and birdlike lungs living in a tropical climate be like? Of what would such an animal be capable? If a full complement of dinosaurs with these metabolic characteristics were released into tropical Africa, who would win out? Would *Tyrannosaurus* and *Triceratops* be ousted by the present incumbents of the world's terrestrial ecosystems, or would the terrible lizards be just too much for mammals to handle? We may never know exactly how dinosaurs regulated their internal environment, but whatever advantage they had over other Mesozoic animals allowed them to dominate as large terrestrial creatures for over 140 million years. We should be clear about what this means. Dinosaurs were the best at being big on land in the Mesozoic. They enjoyed domination over nondinosaurian reptiles in this niche for an unimaginably long period of time. Even mammals, which so dominate as terrestrial behemoths in our modern world, could not find a single chink in their armor. Dinosaurs stand out as a pinnacle—perhaps *the* pinnacle—in the evolution of large land animals.

Among paleontologists and the public alike, dinosaurs are now finally receiving the respect that 140 million years of unchallenged hegemony deserves. But as human history has repeatedly shown, even the most powerful dynasties do not last forever. The riotous success of dinosaurs has always been overshadowed in the public imagination by one ugly little fact: they became extinct 65 million years ago (except for birds, of course). Dinosaurs died out, mammals survived; what clearer indication of failure and success could there be? Taken as an isolated fact the conclusion seems inevitable, but understanding what happened at the end of the Cretaceous leads to a different interpretation, and one that clearly implicates the success of the dinosaurs in their own demise.

Most paleontologists now suspect that the last dinosaurs of the Cretaceous were wiped out by the aftereffects of an asteroid or a comet that slammed into the Yucatán Peninsula of Mexico. Bits of space junk hit the Earth all the time, but this one was a monster. Perhaps 10 km (6 miles) across, the missile was as wide as San Francisco and traveling at 100,000 km (62,000 miles) per hour. On impact, it punched a hole in the Earth's crust 40 km (25 miles) deep and unleashed an explosion equivalent to 100 million hydrogen bombs. Any animal within 3 to 4000 km (up to 2500 miles) of ground zero would have witnessed a blinding flash and a last few moments of calm before the sky became an inferno. Everything organic within this radius caught fire. Continental-scale wildfires swept the Earth. Soot from the inferno joined millions of tonnes of gas and dust thrown into the atmosphere. The black pall of destruction soon girdled the whole planet and light from the sun was blocked out completely. Temperatures plummeted, photosynthesis stopped, and plants began to die. Heat from the impact split apart molecules of nitrogen and oxygen. These recombined into nitrous oxide and then into concentrated nitric acid, which rained down all over the planet. The dust may have cleared in months, but the impact released huge amounts of gas into the atmosphere, which probably contributed to a brief global winter, followed by a runaway greenhouse effect. It was a new Earth. The dominant large animals of the old order had been powerful, sleek, aggressive, and swift, but the Yucatán impact changed all the rules of the ecological game. The animals destined to inherit the Earth would be those that could crawl through hell.

During periods of extreme environmental crisis, large land animals seem particularly vulnerable. Extinction events at the end of the Permian, Triassic, Jurassic, Cretaceous, and at several points during the Cenozoic hit big species as hard as, or harder than, small ones. Small animals need less food, they can hide in burrows or crevices to avoid extremes of temperature or adverse weather, and they often exist in huge numbers. Abundance is probably a key factor: killing 10,000 elephants is one thing, but 10 million rats? Even with a 99 percent death toll there would be 100,000 rats left to crawl back from the brink, but what hope would there be for 100 elephants? Rats could breed themselves back to relative safety in years, but elephants would require centuries, and for much of this time they would be vulnerable to any other misfortune capable of seeing off a few tens or hundreds of individuals. Large land animals are also handicapped by their dependence on a relatively simple feeding hierarchy: herbivores eat large volumes of plant material and carnivores eat large volumes of herbivores. If photosynthesis stops, the rug is pulled out from under the whole system. Without copious amounts of easily accessible vegetation, the herbivores die and soon the carnivores follow. Small animals, on the other hand, may be able to find enough food in the nooks and crannies of the environment to see them through. Seeds, roots, and tubers can be dug up and eaten. Small insect-eaters subsist on a virtually indestructible food resource that bounces back from environmental disturbance very quickly. The way to survive cosmic catastrophe is to be small, numerous, geographically widespread, and able to eat other natural survivors.

Late-Cretaceous dinosaurs were all large compared with mice and shrews and, as far as we can tell from the remains of their skulls and teeth, arranged within the sort of simple feeding hierarchy that is particularly vulnerable to catastrophic collapse. Herbivorous dinosaurs would have relied on large quantities of easily accessible terrestrial vegetation, and carnivorous dinosaurs, in turn, could not have survived without the herbivores. Once the vegetation died, the whole system toppled like a house of cards. Why did dinosaurs become extinct at the end of the Cretaceous? Probably because of the attributes that had allowed them to dominate so completely as large land animals for the previous 140 million years.

From a geologic perspective the Earth is far from being a peaceful, stable place. Certainly there are long periods of relative tranquillity, but there are also times of extreme crisis, and animals that are supremely adapted to their environments under normal circumstances have no guarantee of being able to ride out the periodic and inevitable storms. There is, perhaps, a lesson here for us all.

The impact at the end of the Cretaceous cleared the land of dinosaurs and skewed the Earth's fauna dramatically toward the small end of the body-size spectrum. The fossil record of birds and mammals is too fragmentary to gauge exactly how they suffered during the closing years of the Mesozoic, but they managed to avoid the dinosaurs' fate. Lizards, snakes, and amphibians came through, and most families of crocodiles and turtles extant in the last stage of the Cretaceous lived to see the dawn of the Cenozoic. But it was a Lilliputian world compared with what had gone before. Mammals scurried around in the undergrowth or scampered along the branches of trees; birds fluttered in the air; snakes struck out from their hiding places to grab furry meals; lizards continued to rid the world of excess invertebrates just as they had always done. The ecological stage was set for the survivors of the Cretaceous catastrophe to expand their horizons. The biosphere has had 65 million years to come to terms with the fall of the mighty Mesozoic giants. It has been an interesting time.

CODA

At least one extremely important paper on dinosaur physiology has been published since this chapter was written. For some time paleontologists have been aware of the existence of an exquisitely preserved theropod dinosaur unearthed in Italy dubbed *Scipionyx samniticus*. Not only is *Scipionyx*'s skeleton virtually complete and still articulated, some of its soft tissues have also been preserved. John Ruben went to Italy to study the fossil with an unusual piece of paleontological equipment—an ultraviolet light. Because of the fluorescent properties of the minerals that had slowly turned *Scipionyx* into rock, the internal organs showed up more clearly under ultraviolet than under visible light. Unfortunately, the heart and lungs did not survive the process of fossilization, but the outline of the liver, according to Ruben, was very clear, possibly because of the residual presence of liver bile pigments. Ruben concluded that *Scipionyx* had a crocodile-like piston ventilation system of the sort that he had previously attributed to the damaged *Sinosauropteryx*. If Ruben is right, then this theropod at least had an internal organization very different from that of birds. What this finding might imply about the theropod-bird link is still being debated, but it does not refute the theory that the latter evolved from the former. In the opinion of the majority of paleontologists the evidence that birds sprouted from somewhere on the theropod family tree is simply overwhelming.

Interestingly, Ruben has changed his tune about what the pistonlike ventilation system might tell us about the metabolism of these dinosaurs. He still maintains that the bulk of the evidence, particularly the lack of respiratory turbinates, indicates low resting metabolic rates in some or all theropods. However, noting that extant lizards and snakes seem to get by perfectly well with a ventilation system based on simple inward and outward movement of the rib cage, he suggests that the piston system of *Scipionyx* may indicate that it was capable of sustaining oxygen-consumption rates, and therefore activity levels, well beyond those of even the most active living reptiles. In other words, *Scipionyx* may have had what might be considered the best of both worlds: a low resting metabolic rate combined with a greatly increased top-end performance, allowing it to rest like a lizard and run like a gazelle.

It is an intriguing thought, but there is at least one problem with the idea. Ruben's own theory about the evolution of warm-bloodedness in the lineage leading to mammals suggests that resting and maximum metabolic rates are linked in some way so that the former increases as a consequence of natural selection favoring increases in the latter (see Chapter 3). The linkage he proposed was that an elevated aerobic capacity entails the expansion of associated metabolic "support functions," leading naturally to increases in resting metabolic rate. If *Scipionyx* had a reptile-like resting metabolism but a mammal-like aerobic capacity, then it must have found some way of sidestepping Ruben's earlier conclusion. How did this dinosaur manage to sustain and fuel a highly active lifestyle yet avoid the sort of increases in metabolic support functions that would have led to a higher resting metabolism? It's difficult to have it both ways. However, although the ratio of maximum to resting metabolic rates is remarkably constant at around 10 to 20 in living tetrapods, a few mammals have ratios as high as 70 (see Jones and Lindstedt, 1993; Paul, 1998). Ruben's suggestion about *Scipionyx*'s metabolism, therefore, while not sitting *too* comfortably with the general thrust of his earlier theory, is not without empirical precedent.

[5]

LIFE ON THE EDGE

The survivors of the devastating impact at the end of the Cretaceous were the seed from which all subsequent tetrapod life arose. The staggering diversity of life on our planet today is a reflection of the ability of these fortunate few to adapt and diversify in the post-Mesozoic world. Different types of animals have varied in their fortunes under different environmental circumstances, and the patterns of relative success and failure in time and space are intimately related to the costs and benefits of running different types of metabolic engines. In later chapters we will explore how the Cretaceous survivors got on with each other, but first we need to consider how they have coped with the shape of the planet.

The Earth is a sphere that orbits the sun, spinning as it goes. The axis of rotation leans over at an angle of 23 degrees, which causes the northern and southern hemispheres to swing alternately toward and away from the sun once in each orbit. The North and South Poles are thus hidden from and then exposed to sunlight once a year, so polar winters are continuously dark and polar summers continuously light. Constant summer sun followed by winter darkness at high latitudes causes dramatic seasonal changes in temperature. Day lengths vary throughout the year at mid-latitudes and stabilize at roughly twelve hours in the tropics, so winter and summer temperatures gradually become more stable toward the equator.

Polar regions are also much colder than the tropics. This has nothing to do with the total duration of dark and light, because 365 day-night cycles at the equator are the equivalent of two six-month cycles at the poles. The poles are cold because sunlight approaches at an oblique angle, travels a long path through the atmosphere (losing energy as it goes), and spreads over a much wider area when it finally makes groundfall. At the highest latitudes, ice and snow also reflect much of the sun's energy straight back out into space, so the total amount of sunlight absorbed by a given area of ground over the course of a year is much less at the poles than at the equator.

Variation in sunshine exerts a profound influence on life. In tropical regions sunlight is strong, temperature and day length are relatively stable, and plants are able to churn out abundant wood, leaves, fruits, and seeds all year round. In contrast, winter is a tough time for plants in temperate regions, and many of them respond by shedding their leaves and shutting down until spring returns. Weaker sunshine and winter dormancy combine to limit the amount of plant material produced in temperate forests to less than half that of their tropical counterparts. Further north on the tundra of North America and Eurasia, sunlight is even weaker, temperatures stay below 0°C (32°F) for most of the year, and when the ground eventually thaws, plants rush through a short and frenetic growing season. The lack of sunlight and the harshness of the tundra climate limit productivity to a level similar to that of many deserts. At the highest latitudes, vegetation cannot cope at all and the landscape becomes barren and mantled with ice and snow. Plants form the base of terrestrial food chains, so the falling off in plant productivity as latitude increases, combined with the severity of the physical environment, sets a limit on the number of herbivores and carnivores that can survive too. Sunshine is the fuel for nearly all Earth's creatures, and differences in the supply of this vital resource ramify through the whole web of life.

The total number of species also declines toward the poles. The reasons for this are complex and not yet fully understood—at least twenty-eight theories have been proposed—but some factors are clear. Species diversity tends to be higher in environments that produce a lot of plant material. In theory, a productive environment could just encourage larger populations of particular plants and animals, but in practice the number of species tends to increase too. Tropical woodlands, for example, accommodate all their plant productivity by becoming tall and vertically complex, with many different layers of vegetation in the canopy. Complexity means nooks and crannies for other types of plants to exploit. The diversity of the forest then feeds on itself as herbivores adapt to the wide range of opportunities and the carnivores evolve in pursuit. The predictability of rain forest production is also important because it provides a constant larder for leaf- and fruit-eating species. Diners on such foodstuffs are much rarer in temperate woodlands because trees shed their edible parts during the winter.

Animals from temperate regions also tend to be tolerant of a greater range of environmental conditions than those from tropical areas because they have to survive the roller coaster of the seasons. Species capable of tolerating environmental changes in time are also likely to be tolerant of environmental differences in space, so they should be able to spread over wider areas. If we imagine the geographic range of a species to be a jigsaw piece, then a temperate community is like a jigsaw with 100 large pieces and a tropical community like a jigsaw with 1000 small pieces: the larger the pieces, the fewer will fit into any given area. This effect is particularly severe on our spherical planet because the area of land at the poles is much smaller than around the bulging girth of the equator. Even if the geographic ranges of animal species were the same size the world over, polar regions could never accommodate as many as tropical ones because there simply isn't room. These are some of the ideas, but the main message from modern research is that global patterns of diversity and their causes are exceedingly complex.

Mammals, birds, amphibians, and reptiles all decline in diversity toward the poles, but the gradient is steepest among the cold-bloods. Reptiles and amphibians occupy the bulging low and mid-latitudes of the Earth in great variety. Frogs, lizards, and snakes outnumber mammals two or three to one in moist tropical forests and by even more in tropical waterways. But head to the poles, and the diversity of cold-bloods declines precipitously and then fizzles out altogether, while mammals and birds continue their march north and south. Northern Norway and the southern tip of South America are the ends of the Earth for land-living amphibians and reptiles, leaving a broad northern belt of the Americas and Eurasia, all the North Atlantic islands including Greenland and Iceland, and the southern landmass of Antarctica completely free of cold-blooded tetrapods.

Cold-blooded metabolic engines simply do not tick over fast enough to keep their owners warm in such extreme northern and southern climates. Large cold-bloods in moderate climates may benefit from the relative temperature stability that big bodies provide, but just being big is not a viable strategy at the poles. A large reptile would have to bask in the sun to warm up and then hibernate or shelter from the cold in the depths of winter, and as animals get bigger, both types of strategy become more difficult. Large animals warm up slowly in the sun because of their relatively small surfaces, and they have difficulty finding holes big enough to hide in. Two-tonne animals of all sorts have to stand out in the open and brave whatever the environment throws at them. A period of sustained cold would eventually reduce the internal temperature of an exposed big reptile to that of the environment, with all the associated reductions in the rates of physiological processes and athletic prowess. Below 20°C (68°F) a big reptile would no longer be able to digest the food in its stomach. Much colder and it would simply become a landmark. At sub-zero temperatures, ice crystals grow inside cells, rapidly damaging them beyond repair. Some small coldbloods escape periods of freezing by hiding or hibernating, and a few even produce antifreeze to protect their delicate cellular machinery, but surface temperatures close to the poles in winter would spell death for any exposed amphibian or reptile. To survive in an environment that fluctuates between 20°C and minus 60°C (68°F and minus 76°F), tetrapods need an onboard heating system at least and some additional countermeasures besides.

Warm-blooded animals have a number of strategies to cope with cold climates, but they are all aimed, at least initially, at preventing shivering. Shivering uses up huge amounts of energy, so animals go to great behavioral lengths to avoid doing it. Naked humans begin to shiver at around 28°C (82°F), and below this point our metabolic rate doubles with each 10°C (18°F) drop in temperature. Most tropical mammals have shiver-points between 20°C and 30°C (68°F and 86°F). Arctic lemmings, because they are small and lose heat so readily, begin to shiver at 20°C (68°F), which is why they spend the long Arctic winters under the snow. Huskies, however, can withstand air temperatures of minus 20°C (minus 4°F) without shivering, and Arctic foxes can rest comfortably in the snow at a mind-numbing minus 45°C (minus 49°F). An Arctic fox and a similar-sized human child produce about the same amount of heat in total, so Arctic foxes must lose heat to the environment more slowly. In other words, humans and Arctic foxes differ mainly in the effectiveness of their insulation.

In the depths of winter an Arctic fox's pelt may be 5.5 cm (2.1 inches) thick—remarkably luxuriant for an animal of such a size—and the fur itself is very dense with a fine wool undercoat. Many high-latitude mammals have pelts with the same sort of fine, soft texture, which makes them very popular with indigenous people and, unfortunately, many fashionable temperate humans as well.

Curiously, the fur of polar bears has never been popular for clothing, partly because of the hazards of removing it from its rightful owner but mainly because the hairs are rather coarse. Despite being 7 cm (2.7 inches) thick, polar-bear fur has quite a low insulation value, similar to that of temperate foxes and rabbits. Polar bears are very large, compact animals so they lose heat relatively slowly, but the main reason for their paradoxical fur is probably that they regularly swim in icy polar waters. When a polar bear enters the sea, the insulation value of its fur effectively drops to zero, because cold water replaces the trapped layer of warm air next to the skin. Bears manage to survive immersion in ice water mainly because of a thick layer of fat beneath the skin, which also compensates for the fur's relatively poor insulating qualities on land. When a bear climbs back onto the ice pack, the water drains out of the coarse pelt quickly, to be replaced by a more effective insulating layer of air.

Although a polar bear's hairs are quite coarse, they have an unusual structure that some have claimed improves the pelt's heat-trapping qualities. Each hair is hollow and may act like an optical fiber, channeling short-wavelength radiation through the pelt so that it can be absorbed by the pitch-black skin underneath. But if black surfaces are good at absorbing heat, why don't polar bears just have black fur? While a black pelt might be useful for soaking up the sun, a seal resting by an ice hole would see a black bear coming a mile off, and being warm is scant compensation for being hungry. Polar bears have even been seen stalking seals with snow piled up on their noses, a telling indication of how crucial camouflage is for these animals.

More generally, the thermal characteristics of pelts of different color are quite complex. Desert ravens, for example, have black feathers that soak up heat so effectively that the feather tips on a hot day may reach 80°C (176°F). Very little of this heat reaches the bird's skin, however, because a layer of air trapped in the plumage acts as an insulating shield. But the really neat trick is that the feathers get hotter than the surrounding air; this reverses the thermal gradient between the raven and the environment and encourages heat to flow *out* of its body. Desert ravens may be black because it allows them to thermoregulate more effectively. The coats of ravens and polar bears suggest, against all intuition, that black may be best in hot climates and white in cold ones. Zebras remain a puzzle.

Water sucks heat out of warm objects at a very high rate—which is why blacksmiths dunk hot horseshoes in water rather than wave them around in the air—so warm-blooded marine animals such as seals and whales have to cope with immersion in a medium with phenomenal cooling power. The options for aquatic mammals are to live with a low body temperature, increase metabolic heat production, or have very effective insulation. The body temperature of seals and whales is maintained at 36°C to 38°C (97°F to 100°F), which is typical of mammals as a group, so the first option seems to have been bypassed. The metabolic rate of some seals and porpoises has been measured and turns out to be about twice as high as that of typical mammals of equivalent size, suggesting that turning up the boiler is at least part of the strategy. But even a doubling of metabolic rate is insufficient to counteract the heat-sapping effect of the water in which these animals live, so the main solution has to be insulation, and plenty of it. Some 50 to 60 percent of the cross-sectional area of a harp seal is made up of blubber located in a ring between its skin and internal organs. In cold water, seals cut off blood flow to their skin, leaving the blubber as an insulating shield. Their skin temperatures drop close to that of the surrounding water, which means that there is effectively no gradient for heat to flow along (the same principle is used by terrestrial pigs, some of which can survive Alaskan winters quite happily with only a sparse covering of bristles). Blubber is such an effective insulator that the shiver-point of harp seals is well below 0°C (32°F), so they can rest in even the coldest water without elevating their metabolic rates.

But what about a seal's appendages? Its body may be encircled by blubber, but there is a limit to the amount of insulation that

FIGURE 5.1 Caribou (reindeer). Length 2.2 meters (7.2 feet).

can be carried on skinny protuberances like flippers. The same problem applies to the spindly legs of caribou (North American reindeer; Fig. 5.1), huskies, and Arctic foxes, and the flattened feet of ducks and gulls. If a duck stood on the frozen surface of a lake with warm feet, the ice would melt, its feet would cool, and it would rapidly freeze to the spot. It doesn't matter how effective an animal's insulation is if there are gaping holes through which heat can rush out into the environment.

This problem is solved in all of these animals by an ingenious network of arteries and veins located between the trunk and the appendages. Cold blood coming up from a duck's feet, for instance, is pushed through blood vessels that run very close to others containing warm blood on the way down. The thermal gradient between the two causes heat to flow out of the descending blood before it ever reaches the feet. While a husky's body remains at 38°C the temperature of its paws is often close to zero. The temperature difference between paws and snow is thus minimized so that little heat escapes. When huskies exercise vigorously, as they often do when pulling sleds, the heat exchanger is overwhelmed and their legs warm up; the legs are then used to radiate excess heat in the same way that elephants use their ears.

Keeping legs and feet cold is a very effective way of preventing heat loss, but the strategy has required some rather subtle changes in body chemistry. The physical properties of fat, for example, change dramatically with temperature (think of how the spreadability of butter can vary). Fat in the torso of a caribou behaves just like butter, becoming hard when chilled. If a caribou's legs contained the same sort of fat, its feet and lower leg joints would rapidly seize up. Arctic animals get around this problem by having fat in their feet with shorter hydrocarbon chains and a much lower melting point. Eskimos have always used the marrow fat at the top of a caribou's leg as a solid food and the foot fat as a fluid lubricant. Similarly, some farmers still use fat extracted from the feet of cattle—neat's-foot oil—to keep their leather boots and plough harnesses supple in freezing weather. As is so often the case, science has rediscovered the knowledge that indigenous people have long taken for granted.

These are the main ways in which polar warm-bloods cope with very low temperatures. Large bodies supplemented with effective internal or external insulation and strategically located heat exchangers allow thermal equilibrium to be maintained most of the time without increasing internal heat production. This strategy makes a lot of sense. Home owners invest in insulation for their boilers, knowing that the outlay will be amply repaid. Although it takes energy and resources to grow and maintain fur, blubber, and heat exchangers, it is clearly a more energy-efficient approach than the sort of brute-force elevation of metabolic rate employed by animals like shrews. No doubt shrews would have better insulation and lower metabolic rates if the long-term energetic benefits of looking like a powder puff outweighed the rather more immediate costs.

The perceptive reader will have noticed a potential problem with these examples of polar adaptation which I can't resist illustrating with an old joke.

Annie the polar bear and her son Pippalook are walking across the ice pack looking for seals. Pippalook seems to be lost in thought. "Mum," he says after a while, "can I ask you a question?"

"Of course, son," says Annie, "what's the problem?"

"Well," says Pippalook, "there's something I've been wondering about for a while." Pippalook bites his lip. "Mum," he says, "am I really a polar bear?"

Annie, somewhat taken aback, looks down at her son, and then at the landscape of ice and snow stretching to all horizons. "Er, well, *I'm* a polar bear, son," she says. "And your father's a polar bear too. And you've got white fur, a little black nose, and we're currently on an ice pack looking to murder a seal. I'm no zoologist, son, but the evidence definitely points to your being a polar bear."

"Oh," says Pippalook despondently, and trudges on wearily through the snow. "But Mum?"

"Hmm?"

"It's just that . . . if I'm really a polar bear . . ."

"Yes?"

". . . then why am I so bloody cold?"

Pippalook's question illustrates an important point. All large animals start out as small ones. Small animals do not have the stature to carry large amounts of fur or fat, and they lose heat rapidly through their relatively large surfaces. If adult polar bears feel the cold, how on earth do their cubs survive?

Polar bears give birth in late December or early January when air temperatures in the Arctic can fall below minus 40°C (minus 40°F). Cubs weigh only 700 grams (1.5 pounds) at birth and they are blind, wet, lacking in fur and fat, and cannot shiver. Such defenseless animals could never survive exposure to the Arctic elements, so pregnant females excavate dens in which they give birth and spend several months caring for their cubs. Warmed by the mothers' body heat, dens are usually close to freezing even if the external temperature falls to minus 40°C. Even at 0°C (32°F), however, it is unlikely that cubs could survive without additional protection. Observations of polar bears in their dens—few and far between for obvious reasons—suggest that the female curls up, clasps her back and front paws together, and presses the cubs to her body with her thickly furred legs. She may also warm the infants with her breath. While mother and cubs are locked in this life-giving embrace, the cubs feed regularly on her milk and grow very quickly. After three months they weigh around 10 kg (22 pounds), have a thick layer of fur and unusually high metabolic rates for animals of their size. At this point the cubs are fully equipped to venture out of their dens and brave air temperatures as low as minus 30°C (minus 22°F).

Caribou (Fig. 5.1) and musk oxen (Fig. 5.2) do not dig dens, so their young have to cope with the tundra climate right from birth. Calves of both species weigh around 6 kg (13 pounds), much heavier than polar-bear cubs but still tiny relative to their parents. Caribou calves are active soon after birth and seem to have excellent temperature control. They benefit in particular from a pelt of air-filled hairs that provides good insulation so long as it remains dry. When cold winds and rain batter the breeding grounds, caribou calves can elevate their rates of heat production by a factor of five, but even with these thermoregulatory abilities they are particularly vulnerable to the combined effects of wind and rain, so mortality tends to be very high in years of bad weather.

Unlike caribou, which migrate north during the summer, musk oxen are resident all year round on the tundra of Greenland, Canada, Alaska, Norway, and Svalbard. Calves are well insulated with long fur that, unusually for mammals, extends down to cover their spindly

FIGURE 5.2 Musk ox. Length 2.2 meters (7.2 feet).

legs. Musk oxen are also born with vast deposits of brown adipose tissue in their abdominal cavities, the sole function of which is to produce heat by metabolic reactions. This tissue allows calves to increase their overall rate of heat production by 50 percent without shivering, to a level thirteen times that of a human at rest. With these adaptations, a newborn calf can immediately raise its body temperature to 75°C (135°F) above that of its surroundings.

Even at birth, polar bears, reindeer, and musk oxen are huge compared with some animals that brave the Arctic climate. Lemmings, for example, give birth to litters of tiny offspring beneath the snow, often in midwinter. The icy roof of the nursery affords the same sort of protection from the elements as a polar bear's den, and huddling with the mother, and with each other when she is absent, further reduces the rate of heat loss. Even so, baby lemmings are just too small to maintain a normal mammalian body temperature, so for the first ten days or so of life they are essentially cold-blooded. They also exhibit an extraordinary, but still poorly understood, tolerance of cold. Nestlings chilled to 3°C (37°F) have been revived without any apparent ill effects. Over the first couple of weeks of life, lemmings gain the ability to shiver, grow thick fur, and lay down deposits of heat-producing fat. A very similar pattern of tolerance to hypothermia and a gradual transition from an essentially cold-blooded metabolic regime to a warm-blooded one is characteristic of many small Arctic species, including collared lemmings, ermine, and most altricial birds (those that give birth to helpless young) such as snowy owls, ravens, crossbills, and snow buntings.

The Arctic is a harsh place to live and breed, but it is positively balmy compared with the polar region at the other end of the Earth. The average winter temperature in the Antarctic is minus 60°C (minus 76°F), and even the warmest summer days are colder than most winter nights at the North Pole. The difference between the two regions stems from the fact that the Antarctic is a landmass capped with ice, while the Arctic is essentially a frozen ocean. The surface of the Arctic is never more than a few meters above sea level, and the liquid water below the ice provides a constant reservoir of heat. The Antarctic, in contrast, with an average altitude of 2300 meters (1.4 miles), is three times higher than any other continent, and each 100 meters (110 yards) of altitude represent a drop in temperature of 1°C (1.8°F). The Arctic is also warmed by high-pressure weather systems that spill northward from the land areas of North America and Eurasia, but Antarctica is isolated in the great southern ocean, 2500 km (1550 miles) from Australia and 4000 km (2480 miles) from Africa. Low-pressure systems isolate Antarctica from the rest of the world's weather and keep temperatures well below zero throughout the year.

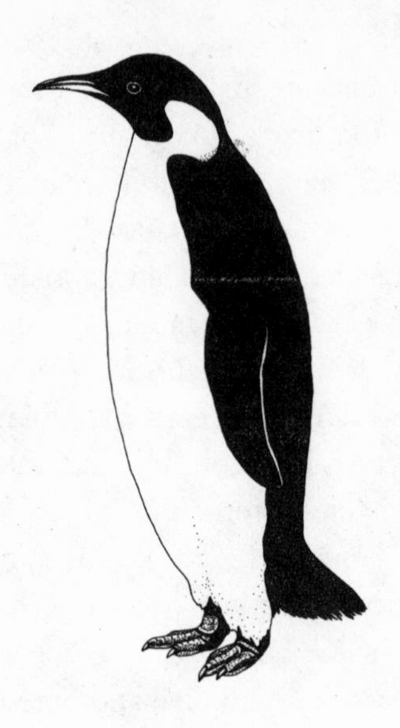

FIGURE 5.3 Emperor penguin. Height 1.2 meters (3.9 feet).

Antarctica's harsh climate and isolation have resulted in a relatively impoverished fauna. Forty species of land mammal frequent the Arctic, but none has managed to colonize the southern pole. The only Antarctic mammals are seafaring whales and seals, the latter escaping to the relative warmth of the sea in winter. There are also eight resident species of bird in the Arctic, but none has managed to make a home in the Antarctic wilderness. Of the species that do exploit this great southern landmass, the most remarkable must be the emperor penguin (Fig. 5.3). Emperors are the largest of all penguins, standing a regal meter (39 inches) tall and weighing 30 to 40 kg (66 to 88 pounds). Many birds breed in Antarctica, but only the emperor does so in the winter. Penguins cannot fly, and 30-kg birds are too heavy to fly in any case, so emperors walk as much as 100 km (62 miles) across snow and ice just to reach their breeding grounds. They are not really accomplished walkers either, with their short legs and webbed feet, so the whole migration takes place at an achingly slow pace. Often they walk in single file to shelter from the biting Antarctic winds. Each bird presses up close to the one in front and keeps in step, so the whole chain snakes forward across the freezing ice in what can only be described as a synchronized waddle. When they eventually arrive at their breeding ground, which looks much the same to human eyes as most of the places through which they have just spent days walking, the female lays a single large egg. The male picks it up, balances it on his feet, and covers it with a warm fold of skin from his abdomen. The females then turn around and slowly walk all the way back to where they started.

When the females arrive at the ice edge, they take to the water and feed as fast as possible to build up the fat reserves depleted by their 200-km (124-mile) round trip. Meanwhile the males stand in total darkness with the next generation of emperors balanced precariously on their feet. There is no food for penguins on land, so the males starve throughout the nine-week period of incubation. Feathers and fat offer excellent insulation, but the shiver-point of emperors is only around minus 10°C (14°F). Heat exchangers limit heat loss from the legs and feet, but the males probably release a fair amount to incubate their eggs. As the temperature dips to minus 60°C (minus 76°F) and winds of up to 300 km (186 miles) per hour scour the landscape, some other method of staying warm is clearly needed. The crucial adaptation seems to be huddling, a behavior that reduces the surface area of each bird exposed to the steep temperature gradient and freezing winds. Isolated birds lose their fat reserves at double the rate of huddlers, and as males on average lose 40 percent of their body mass during incubation, huddling can clearly make the difference between life and death.

After two months of separation, the females make the long waddle back to their eggs, which are beginning to hatch, and when they arrive, the starving males embark on the long trek back to the ocean to feed. The females keep the chicks in their belly pouches for three weeks or so, feeding them on demand by regurgitation from the crop. Then the males return to the breeding ground again. After seven weeks the chicks weigh about 2 kg (4.4 pounds) and are too large for their parents' brood pouches. At this stage they are covered with a thick grayish coat of downy feathers that affords excellent insulation, but they still have to huddle like their parents to have any hope of survival. Meanwhile the adults trek back and forth to the edge of the ice to refill their bellies. Over the course of a breeding season each pair may make as many as fourteen trips. By five months the chicks weigh 15 kg (33 pounds) and their parents leave them for the last time to return to the sea. A week later, driven by hunger, the chicks follow in their footsteps to the ice edge and their first glimpse of the sea. Perhaps not surprisingly, the mortality of emperor chicks is depressingly high. Only 20 percent make it through their first year, compared with around 80 percent for summer breeders like the king penguin. Why such an arduous and hazardous pattern of breeding behavior should have evolved in the most extreme climate on Earth remains a mystery; one for future generations of cold-hardy biologists to ponder and perhaps solve.

At the other end of the terrestrial temperature scale, deserts pose a different set of problems and opportunities. Unlike polar regions, deserts are home to a rich diversity of cold-blooded animals. Most are invertebrates and reptiles of one sort or another, but even some moist-skinned amphibians have evolved opportunist lifestyles that allow them to eke out a living under extremely arid conditions. Spadefoot toads are perhaps the most remarkable desert-adapted amphibians of all, living out their lives in the searing crucible of the Arizona desert. After the rains, just before the ground begins to bake hard, spadefoots entomb themselves in mucus-lined chambers a foot or so below the surface. Their skins harden into leathery cocoons and they enter what is essentially a state of suspended animation. When the next cloudburst arrives, the males emerge from their chambers, congregate in the nearest pools, and begin croaking to guide the emerging females to a brief, frenzied communal orgy. Having sated their reproductive urges, the adults feed as fast as possible to replenish their food reserves before digging themselves back into the mud and turning out the lights. The eggs begin to hatch in temporary pools within twenty-four hours, and the tadpoles immediately begin to feed on algae, fairy shrimps, and each other. Lizards, birds, and snakes take a heavy toll on the tadpoles and toadlets, but many reach the relative safety of hiding places in muddy cracks and under rocks, where they feed for a few weeks on virtually anything they can find before digging themselves in beside their parents to ride out the dry season as shriveled, hardened bags of potential. When the rains come, the cycle starts all over again.

Although some amphibians have managed to adapt to extreme aridity and high temperatures, they are a minor part of the cold-blooded fauna of hot deserts. It is the reptile clan of tetrapods, particularly lizards and snakes, that dominates in arid climates. Reptiles have relatively impervious skins, and they do not need to lay their eggs in water, so they are at a distinct advantage over their amphibian cousins in hot and dry environments. And unlike mammals and birds, reptiles rely on external sources of heat to warm their bodies, and heat is one commodity that deserts have in abundance. But the physiology of lizards and snakes does impose one fundamental limitation on their adaptative scope. Because they are acutely sensitive to the temperature of their surroundings, reptiles have no choice but to be small enough to hide when temperatures soar or plummet. Small desert mammals face a similar problem: because they cannot afford to lose water by sweating, they must also be accomplished hiders. But some warm-blooded animals have been able to adapt to long periods of exposure to the full force of the desert sun, allowing them to ditch the hiding strategy altogether and grow big.

Perhaps the most familiar large desert animal is the camel. Unusually, camels can cope with body temperatures higher than 38°C. When a camel has access to water, its internal temperature is typically mammalian, but when dehydrated its temperature begins to fluctuate markedly, reaching 34°C (93°F) at night and 41°C (106°F) during the hottest part of the day. A 7°C (13°F) rise in body temperature for a 500-kg (1100-pound) animal adds up to around 3000 kilocalories of stored heat which, if eliminated by sweating, would use up 5 liters (4.5 quarts) of water. So by storing heat during the day and allowing it to escape naturally into the environment at night, a camel prevents itself from dehydrating further. And maintaining a body temperature of 41°C during the heat of the day has another useful effect: because body and air are closer in temperature than they would be if the camel stayed at 38°C, the temperature contrast between the animal and its surroundings is reduced. This slows the rate of heat gain from the environment, a trick at least as important as daytime heat storage for conserving water. Camels do sweat to prevent their bodies from becoming hotter than 41°C, so dehydration is still potentially a problem, but camels can withstand levels of water loss that would be fatal to most other animals. A 10 percent reduction of body weight in water is close to the lethal limit for humans, but camels can suffer a 25 percent loss of weight through sweating without any apparent ill effects. This allows them to survive six to eight days without drinking under conditions that would kill a human within twenty-four hours. Camels also limit heat gain, and thus water loss, with their fur. Given the thick pelts of Arctic foxes and polar bears and the use to which they put them, thick fur might seem a bit out of place in a desert, but a camel uses its pelt as a thermal shield, reflecting sunlight straight back into the environment. The thermoregulatory benefit of a camel's fur is easily demonstrated by cutting it off, whereupon the poor animal is forced to produce 50 percent more sweat in order to rid excess heat from its body.

There is no truth in the legend that camels fill their humps with

FIGURE 5.4 The Saharan scimitar-horned oryx and its straight-horned Arabian cousin are perhaps the ultimate large mammalian desert specialists. Length around 2 meters (6.6 feet).

water before a long journey or otherwise store water in any unusual way, but they can drink as much as 30 percent of their body weight in a single sitting, so their basic strategy is to fill up with water and then make it last as long as possible. Some desert species, however, make this sort of rationing strategy seem rather crude. Perhaps the most remarkable is the oryx (Fig. 5.4), an irascible antelope with a suite of elegant adaptations to searingly hot and dry environments. Oryx make little attempt to hide from the sun, and local people have long maintained that they never drink (although we now know that they drink occasionally, and females seek out waterholes when in calf). How can a mammal possibly survive in the middle of the Sahara Desert without seeking shade and practically without water?

Heat storage is the principal mechanism limiting water loss in oryx just as in camels, but oryx store heat in their bodies at levels that would kill a camel stone dead. As far as I am aware, the harshest conditions to which a dehydrated oryx has been subjected in a climate-control chamber is 45°C (113°F) for eight hours. The animal simply allowed its internal temperature to rise slightly above 45°C for the entire length of the experiment. We still do not understand exactly how oryx achieve this level of heat storage without the sort of disruption to physiological processes and cellular machinery that would kill most other animals, but we do understand the ingenious way in which they protect the most delicate part of their anatomy.

Brains are immensely complex structures that have a nasty tendency to go wrong if heated above normal operating temperature, and 45°C definitely qualifies as too hot. Oryx prevent their brains from scrambling with a cranial refrigeration system that works in the opposite way from that used by polar animals to prevent heat from escaping from their extremities. Blood from an oryx's heart passes through a region in the head called the cavernous sinus, where the carotid artery splits into hundreds of smaller vessels. The other part of the sinus consists of a network of veins emanating from the nasal passages; the blood in these veins is pre-cooled by the flow of air in the nose so that it is colder than the arterial blood on its way to the brain. The difference in temperature between the two networks causes heat to flow out of the arterial blood and protects the brain from overheating.

Like camels, oryx dump most of their excess heat by radiating it back to the environment at night. Nocturnal heat-dumping saves a lot of water, but oryx must use some to dilute their urine and soften their feces, so they are still at risk from dehydration (oryx produce exceptionally concentrated urine, but it is still water-based). The only moisture available to an oryx in the absence of drinking water is from the food it eats. All plants contain some free water, and this can simply be extracted as the material is mashed in the mouth and stomach. The vegetation favored by oryx often contains as little as 1 percent water by weight during the day, but this can increase twentyfold at night as the temperature falls, humidity rises, and desert plants scavenge water directly from the air, so by eating at night oryx exploit the survival tactics of their food plants to rehydrate themselves.

Water is also one of the main products of metabolic reactions—carbohydrates and oxygen react together to give carbon dioxide and water—so all animals have an indirect way of getting moisture from their food. However, while metabolic reactions produce water, they also require oxygen, and water is lost from an animal's lungs every time it exhales. Under normal circumstances, the loss of water associated with breathing equals or exceeds whatever is gained from metabolic processes, so an increased rate of metabolism is unlikely to help an animal balance its water budget. The amazing oryx, however, has even found a way of squeezing excess water out of its metabolic reactions. At night when they are radiating stored heat back into the environment, and while the air is relatively humid, oryx begin to inhale and exhale very slowly and very deeply. Deep breaths allow more oxygen to be extracted from each lungful of air and more metabolic water to be manufactured. Provided the relative humidity of the air stays above 70 percent for most of the night, a deep-breathing oryx will end up with more water inside its body at dawn than it had the previous evening. The production of concentrated urine, daytime heat storage and nocturnal eating, heat-dumping, and deep breathing allow oryx to pull off the astonishing feat of surviving the hottest desert environments without having to drink.

Tales of desert and tundra animals and their extraordinary lifestyles could fill a whole book, but the foregoing examples illustrate the main ways mammals and birds survive the rigors of extreme climates. Perhaps the most intriguing thing about high-latitude and desert species is the way in which they employ similar physical and biological principles for opposite purposes. Polar bears and emperor penguins deploy their relatively small surfaces against the chill to keep heat inside their bodies, while camels use the same principle to keep it out. Small animals hide under the snow to escape the cold of the tundra and under rocks and in burrows to escape the heat of deserts. Fur keeps a musk ox warm and a camel cool. Caribou exploit their large size and ranging ability to move long distances in search of food and to escape the coldest weather, while camels and many other large desert animals do the same to exploit ephemeral water and food. Arctic foxes and seals use networks of blood vessels to stop heat leaking into the environment from their skinny appendages, while oryx use the same system to prevent heat from leaking into their brains. Mammals and birds are famed for their adaptability, and nowhere is this more evident than at the thermal ends of the Earth.

Of all the similarities between high-latitude and desert animals, one general pattern stands out: the denizens of extreme climatic regions tend either to be small enough to hide or big enough to stand exposed to the elements and tough it out. Deserts are home to countless small lizards, snakes, and rodents hiding in cracks and crevices or self-constructed holes, as well as numerous large animals like camels, oryx, elands, dik-diks, addaxes, steenboks, Nubian ibex, plains gazelles of various sorts, zebras, ostriches, wild horses, asses, pronghorn antelopes, kangaroos, leopards, lions, hyenas, dingos, assorted foxes, and some rare coyotes, cougars, lynx, and wolves. But medium-sized animals are noticeably rare and nearly always dependent on burrows when they are present. The jackrabbit (Fig. 5.5) from the desert regions of the southwestern United States is an interesting exception which really just goes to prove the rule.[1] Jackrabbits do not burrow, are too large to wedge

1. The old maxim "The exception proves the rule" has become nonsensical because we now invariably associate the word *prove* with the idea of confirmation (e.g., math-

FIGURE 5.5 Why do jackrabbits have big ears? For the same reason elephants do: to increase the rate at which they can dump heat from their bodies into the environment.

themselves into cracks or under rocks and too small to be heat storers. The lengths to which these animals go to avoid the crippling heat of the day says much about why medium-sized, nonburrowing animals in deserts are so rare. They feed only at dawn and dusk when the chance of heat stress is at a minimum. At midday the desert surface may reach an intolerable 70°C (158°F), so jackrabbits seek out what little shade is available behind rocks, ledges, and mesquite bushes. They favor shallow, shaded depressions on the northern side of obstacles because heat scattered from the surrounding landscape passes over the lip of the depression and thus over the animal's head. While crouched in their hollows, they manipulate their

---

ematical proof). But *prove* can also mean *test* (as in proving ground), and this was the originally intended meaning. "The exception tests the rule" makes eminent sense.

enormous, blood-filled ears to radiate heat back to the relatively cool north sky. One-third of a jackrabbit's metabolic heat can be dumped through its ears without wasting any precious water. In a searingly hot and bone-dry environment, precise sheltering of this kind and anatomic structures capable of dissipating a lot of heat without wasting water are basic adaptations without which a medium-sized, surface-dwelling animal simply could not survive.

At the other extreme, the tundra regions of North America and Eurasia are home to millions of lemmings and voles hiding beneath the insulating mantle of snow. Tiny Arctic redpolls and Siberian tits brave temperatures down to minus 30°C (minus 22°F), but with their very dense blanket of feathers—the best insulation in the animal kingdom—these birds are clearly just hiders of a different sort. And large animals at high latitudes tend to be noticeably so: polar bears, gray wolves, and emperor penguins are the largest of their respective kinds, musk oxen are among the largest members of the ox family (which includes antelopes, cattle, sheep, and goats), and caribou are the second-largest members of the deer family (after moose, themselves no strangers to ice and snow). Arctic foxes are no bigger than their temperate counterparts, but they have reduced their surface areas with stocky bodies, short legs, snub noses, and tiny ears. Many large cold-climate species exploit both the hiding and big-body strategies at different times in their development. Polar bears and emperor penguins are nurtured by their attentive parents in dens or brood pouches when small and only emerge when they have grown big enough to tough it out on their own. Rock ptarmigan, willow grouse, hazel grouse, spruce grouse, and black-billed capercaillie are large and rotund for birds and insulated well enough to survive the tundra and frozen forests of Siberia, but when temperatures plummet in the winter, even these hardy animals dig tunnels in the snow and hide. Arctic hares do not burrow, but, like their desert counterparts, they are highly accomplished shelterers. Although large for hares at 4 to 6 kg (9 to 13 pounds) and unusually itinerant, they are still small enough to find hiding places in hollows and under vegetation and ledges of rock. In winter when deep snow flattens the landscape, they move to higher, steeper ground to find shelter among rocky outcrops and scree slopes.

The predominance of large and small animals is peculiar to deserts and high latitudes. Globally, a tenfold reduction in body length equates roughly to a hundredfold increase in the number of species, and an even greater increase in the number of individuals. But extreme climatic environments favor animals small enough to shelter and large enough to exploit the temperature stability, insulation, and ranging ability that big bodies provide. So in some respects, cold environments and deserts are similar by virtue of being extreme. Both are characterized by temperatures that frequently differ from the normal set points of warm-blooded animals—or the preferred temperature of cold-blooded ones—to a considerable degree. The temperature-sensitive internal machinery of animals must be protected whether the environment is too cold or too hot, and there are predominantly two ways in which this can be achieved. The coldest and hottest lands on Earth are home to the small and the large, the homemakers and the wanderers, the hiders and the striders, and not much in between.

The other striking aspect of extreme climate terrestrial faunas is that only warm-bloods grow to any great size. In very hot and very cold places only mammals and birds, with their powerful metabolic engines, adaptable insulation, and temperature-control mechanisms, can grow bodies large enough to brave constant exposure to the elements. But if it is just extreme climates that confine cold-blooded animals to small bodies, then where are all the 2-tonne lizards and frogs in the warm and equable tropics? Cold-blooded terrestrial animals are relatively diminutive the world over, so extremes of temperature variation cannot be the only factor restricting their adaptive scope. Cold-bloods do not always have to hide from the weather, so why are they all so small?

[6]

WHERE THERE BE DRAGONS

From the frozen wastes of the Arctic to the deserts, savannas, and tropical woodlands of Africa and the Americas, nearly all of the large land animals on Earth are mammals. Polar bears, musk oxen, caribou, and wolves are the most conspicuous large animals in the northern tundra regions of North America and Eurasia. The coniferous forests to the south host wolves, bears, lynx, and more species of deer than any other type of habitat. Temperate grasslands support, or did until recently, deer, bison, wild horses and asses, and saiga and pronghorn antelopes. Temperate forests are home to bears, red deer, wild cats, wolves, and wild pigs. Regions with mild winters and hot summers, such as the areas surrounding the Mediterranean Sea, the southern tip of Africa, southern Australia, and California, are prowled by dogs of various sorts, and grazed by red deer and wild boar in Europe, duikers in South Africa, guanacos in Chile, kangaroos in Australia, and mule deer in California.[1] Camels, oryx, gazelles, foxes, and hyenas roam deserts, while yet more gazelles, elands, pumas, leopards, and cheetahs occupy their scrubby margins. On the savannas of Africa the mammalian hordes reach their greatest

---

1. Duikers are small antelope-like mammals. Guanaco are llama-like members of the camel family.

FIGURE 6.1 The okapi, a member of the giraffe family from the rain forests of the Congo. Head-body length up to 2 meters (6.6 feet).

prominence with a bewildering variety of antelopes, zebras, buffaloes, giraffes, elephants, big cats, hyenas, and hunting dogs. Okapis (Fig. 6.1) and elephants browse the lower layers of tropical seasonal forests, while the upper stories are home to animals like langur monkeys in India and koalas in Australia. The enormous productivity of leaves, seeds, and fruit in the canopy of rain forests has given rise to a high diversity of flying and tree-dwelling groups such as bats, sloths, and primates, along with a rather restricted ground fauna of anteaters, peccaries, tapirs, and capybaras (Fig. 6.2). Primates are particularly diverse in rain forests, with different types of monkeys in the New and Old World, lemurs in Madagascar, gorillas and chimpanzees in Africa, and orangutans, gibbons, and tarsiers in Asia (Fig. 6.3). On the continental land areas of our planet, the niche of large land

FIGURE 6.2 (a) Giant anteater (head-body length 1.1 meters [3.6 feet]); (b) peccary (length 1 meter [3.3 feet]).

animals is almost totally occupied by mammals. Species of reptiles outnumber mammals by almost two to one globally, and it has probably been this way for much of the last 65 million years, but our scaly cousins have contributed conspicuously little to the world's terrestrial megafauna. Why?

Some researchers believe that reptiles are simply incapable of evolving really big bodies on land because of the limitations of their metabolic engines. The paleontologist and artist Gregory

FIGURE 6.2 *(continued)* (c) tapir (length 2 meters [6.6 feet]); (d) capybara, the largest member of the rodent family (length 1.2 meters [3.9 feet]).

Paul maintains that terrestrial reptiles[2] could never grow as large as elephants because they would be unable to support and move their enormous bodies around in the Earth's gravitational field. As we've already seen, the muscles of reptiles are capable of producing short bursts of power through non-oxygen-based metabolic reactions, but these same muscles, Paul argues, would be incapable

---

2. Paul does not count dinosaurs as typical reptiles. I assume the same here for the sake of argument.

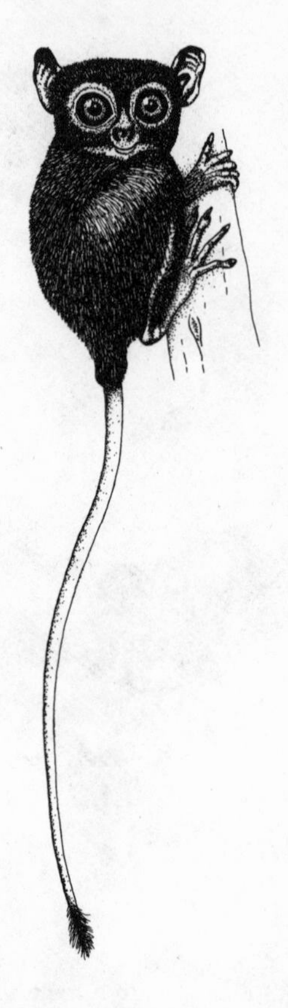

FIGURE 6.3 Tarsier. Head–body length around 14 cm (5.5 inches).

of sustaining the levels of aerobic power required to carry a 5-tonne body around the whole time (not without frequent episodes of belly-resting, anyway). As body weight increases above the level that reptiles with typical cold-blooded engines can support, their muscles would need to increase in size and aerobic capacity to carry the load. Such changes would require larger hearts and lungs to deliver enough oxygen to keep their muscles fired up, and probably an increase in the size of other internal organs performing metabolic support functions as well. In other words, the more gigantic a terrestrial reptile becomes, the more powerful its metabolism has to be, so beyond a certain point it would not be a reptile in the sense that we understand the term. Paul goes on to extend this line of argument to giant dinosaurs with their huge muscles, fast gaits, alleged migratory behavior, and poor design for belly-resting (for example, straight legs rather than sprawled) and concludes that they probably had aerobic capabilities far in excess of living reptiles.

More research is required to evaluate Paul's theory and to estimate the size limits involved—if, indeed, they exist at all—but the line of reasoning is certainly plausible. Paul's "terramegathermy" idea applies only to terrestrial animals, however, because in water the effects of gravity are removed, so there is nothing to prevent fish and aquatic reptiles from growing to much larger sizes.

The size of terrestrial reptiles may also be limited by their relatively slow rates of growth. Most mammals and birds reach adult weight very quickly, usually aided by the considerable efforts of doting, protective parents. Reptiles may grow fast in captivity if they are kept warm and if ample food is placed within easy reach, but low activity levels and lack of parental provisioning in the wild may prevent them from garnering enough food to grow at anything like comparable rates. The world is a dangerous place, and the odds of falling victim to disease, predation, or misfortune increase in absolute terms with age, so reptiles simply may not be able to grow fast enough to reach megaweights before succumbing to one or other of the thousand natural shocks that flesh is heir to.

The terramegathermy and slow-growth ideas are both controversial, not least because some paleontologists believe that dinosaurs had typically reptilian metabolic engines, and some of these creatures grew to over 50 tonnes, but there is no need to wade through all the counterarguments. Even accepting 1 or 2 tonnes as an upper limit for land-living reptiles, animals of such a size would fall into the category of megafauna by any reasonable definition. So where are all the 2-tonne lizards, tortoises, and snakes in our modern world? There aren't any. Again, what is it that keeps our scaly cousins so small?

One of the main factors seems to revolve around the extent to which cold-blooded animals are constrained by the geometry of terrestrial environments. In general, the smaller an animal is, the more complicated the environment will appear from its point of view. Shrews, bats, moths, frogs, lizards, and birds in tropical woodlands play the ecological game of hide-and-seek in an immensely complex three-dimensional space. There are holes in trees, different types of leaves and branches in the canopy, gnarled bark, tangled masses of roots, spaces under fallen logs, cavities in the soil, and so on. Even leaves on the forest floor form an environment of sufficient complexity to hide a small lizard. Grasslands, swamps, and deserts are less complex than tropical woodlands, but all would appear more complex to a lizard than to a horse. A complex environment offers many hiding places, and hiding allows animals to avoid predation, adopt sit-and-wait hunting strategies, and sidestep competitive interactions with other animals. But there aren't many holes in the environment large enough to hide a bear, and fewer still a hippo. Excavating an adequate burrow is easier for a rabbit than for an elephant, and climbing is more energetically demanding for a leopard than for a squirrel. Given the difficulties of finding suitably large hiding places and the problems associated with climbing and burrowing, it is perhaps not surprising that really big animals tend to abandon the strategy of complete concealment and do without the third dimension of height and depth altogether. In other words, the ecological theater for really big creatures consists largely of one environmental plane: the ground. Here predators stalk and chase, prey animals watch and flee, and that's about as sophisticated as it gets.

So the ecological stage for large animals tends to be relatively simple, and this is one of the reasons why scientists interested in unraveling the broad historical and biogeographic patterns of life tend to favor them as objects of study. The skulls of shrews and lizards may turn up side by side in a fossil deposit, but how can we reconstruct their ecological relationship with any certainty? The lizard may have preyed on the shrew, or they may have competed for invertebrate prey, or they may have spent all of their lives in the treetops and leaf litter respectively and never interacted at all. But if the bones of a 1-tonne grazer and an 800-kg (1760-pound) predator are found in the same deposit, then provided we can be sure that the bones were not transported to the site of fossilization from different environments, the interpretation of their ecological relationship is much more straightforward. Both animals must have lumbered around on the ground, often within plain sight of each other. The herbivore would have eaten the vegetation within reach of its neck, and the carnivore most likely ate the herbivore, either by killing it or by scavenging its carcass. When the two animals interacted directly it would probably have been a fight, a chase, or a standoff, with aggressive or defensive behavior involving some combination of speed, strength, ferocity, or armor.

And just as ecological relationships are easier to predict at large body sizes, so too are the advantages and disadvantages of different types of metabolic engine. Judging by the plethora of small warm-bloods and cold-bloods around today, diminutive animals with high-powered and low-powered metabolic systems seem about equally successful, at least at body weights above 2 grams (.07 ounce). Energy-efficient lizards and snakes and gas-guzzling mice and shrews crawl, slither, or scurry around the nooks, crannies, basements, and attics of complex environments, hide from the weather or their enemies, surprise or pounce on their victims, and generally endeavor to keep out of each others' ecological way. Small cold-bloods deposit tiny eggs in secret places to ensure that the environment is populated with offspring in the next generation, while small mammals seek out suitable hiding places to bear and nurture their hungry, demanding broods. In complex environments there are usually places where small animals of all kinds can play to their strengths, hide from their weaknesses, and eke out a living of some sort.

But for large animals limited to the two-dimensional plane of terra firma, the benefits of a fired-up metabolism seem heavily to outweigh the costs. Granted, hide-creep-and-pounce tactics have been perfected by some large animals—tigers, jaguars, and their prey animals in low-latitude forests, for example—but most of the Earth's dry surface is covered with vegetation in which bulky creatures simply cannot conceal themselves completely at all times. In the unforgiving arena of open terrestrial environments, the ecological spoils tend to fall to those with distinctly in-your-face adaptations, namely surpassing swiftness, overwhelming power, bullish ferocity, and ruthless unshakability. Speed, endurance, intransigence, brute force, and all at a moment's notice—these seem to be the key weapons in the arsenal for large earthbound creatures.

So the rules of the ecological game at large body sizes are likely to be particularly tough on cold-bloods. A large lizard could use its metabolic supercharger to sprint as fast as a lion, but reptiles run out of steam very quickly. If a 200-kg (440-pound) lizard could get within a few meters (or yards) of a zebra it might be able to run it down, but getting this close would be as much of a problem for a lizard as it is for any other large predator. Big bodies are easier to see than small ones, make more noise as they attempt to sneak through the undergrowth, and are, all other things being equal, smellier. Zebras are acutely sensitive to all these stimuli and are constantly on the lookout, listen-out, and smell-out for anything untoward. If a zebra caught wind of its reptilian adversary while the latter was still some way off, the zebra would be able to engage its high-capacity aspirated engine and simply walk away. If the lizard made a dash for it, the zebra would only have to keep running until the reptile was exhausted and then resume its nonchalant walk.

But when zebras face lions, their predicament is far more serious. Several lionesses will typically stalk a group of zebras cooperatively. Usually the zebras become aware of what's going on well before the final strike and wisely make a run for it. The chase often involves an extended flurry of activity, with the prey animals milling and scattering and the lions moving fast across the ground, but noticeably not at full tilt. The lionesses run alongside the fleeing animals, maneuver themselves into position, lunge into the milling crowd, separate their chosen target from the rest of the herd, and then one or two lionesses usually move in at full speed to finish off the job. This final act of savagery reveals the full force of a lion's supercharger, but it takes a lot of sustained aerobic activity to set up the kill in the first place.

Lions, like all big cats, are ultimately sprint-killers, outpacing their victims over relatively short distances, and as such they probably approximate the hunting tactics of a hypothetical giant savanna lizard most closely. In animals like wolves and hyenas, however, we see the purest form of the warm-blooded metabolic system, a setup capable of propelling animals around the environment below sprinting pace, but well above walking pace, for *hours*. Once a pack of dogs sets its collective mind on a particular target, the chase usually becomes a battle of pure endurance. Sometimes the victim's will is stronger or the pack's patience inadequate for the task, but all too often the last breath of an animal pursued by dogs is a prostrate, exhausted gasp. The endurance exhibited by mammalian pursuit predators is unquestionably beyond the scope of large cold-blooded animals and always has been, so this particular way of making a living in the environment will always be the unique preserve of warm-bloods.

Endurance is perhaps the main ecological benefit derived by large mammals from their powerful metabolic engines, but maintaining a high and stable body temperature is also highly advantageous. Warm-blooded animals can hunt, flee, fight, and forage at any time of the day or night, in any season, and in virtually any weather, and because our bodies are always at peak operating temperature, we can respond effectively to threats or opportunities at a moment's notice. The internal temperature of a large reptile living out on the plains would be much less stable, and its active responses and physiological efficiency much more variable. Prolonged cold or rainy weather would cool a reptile's body, reduce the power output of its muscles, and lower its walking and running speeds. It is doubtful whether large reptiles could hold their own in direct competition with large mammals in open terrestrial habitats even under ideal climatic conditions, but on cold mornings, in the winter, or during the rainy season, large cold-bloods would be hopelessly outclassed by similar-sized mammals in almost every important energetic regard.

At least, that is the theory. But what about those large reptiles that do manage to make a living in terrestrial environments? There aren't very many of them, and none rival hippos, rhinos, and elephants in size, but we do share our world with a few reasonably big ones. In order to test the rule that large warm-bloods tend to have the competitive edge over large cold-bloods in open terrestrial environments, we need to look in more detail at the ecology and biogeography of our current crop of reptilian heavyweights.

One particularly fascinating and fearsome large reptile can be found on a few islands just east of Bali. Although saddled with the name of the island forming the eastern border of its range, the

Komodo dragon (Fig. 4.7) also inhabits the nearby islands of Rintja, Gili Mota, Padar, and the western end of the much larger island of Flores (and they occasionally turn up on other nearby islands as well). When Governor Steyn van Hensbroek visited Komodo in 1910 to obtain specimens of this animal, he was told by Dutch pearl fishermen that the largest ones reached staggering lengths of 5 to 7 meters (16 to 23 feet). Although van Hensbroek never found dragons anywhere near this size, the 5-to-7-meters claim began to circulate and has been repeated in print by several authors since. Sightings of 5- or 6-meter (16- or 20-foot) dragons were reported in newspapers at the time of the 1934 Surabaja Zoo trip to Komodo, allegedly sanctioned by some of the expedition members. In a *National Geographic* article published in 1936, Lady Broughton wrote that "they [dragons] sometimes weigh 250 lb and attain a length of 12 feet." From 1936 onward, this particular claim was repeated in numerous books and encyclopedias, effectively disseminating the idea of the monster of Komodo to all parts of the globe.

The truth about dragons, however, is a little different. In the late 1930s one researcher decided to ask local people about the size of their legendary neighbor. He placed two sticks 7 meters (23 feet) apart and told the assembled villagers that some Europeans claimed to have seen dragons of such a size. After the people had stopped laughing, one of them picked up the sticks and moved them about 3 meters (10 feet) apart. Contrary to some earlier reports of hysterical natives babbling about 10-meter, 1-tonne monsters,[3] it turns out that local people are not at all prone to exaggeration about the size of their indigenous giant reptile, unlike their occasional European guests. A length of 3 meters, perhaps a little more in extreme cases, is about as large as these animals get. In 1969–71, Walter Auffenberg captured and measured fifty drag-

3. Piazzini (1960).

ons, and the largest twelve measured between 2.25 and 2.6 meters (7.3 and 8.5 feet) in length and weighed, on average, 47 kg (104 pounds). Clearly, the monsters of Komodo are just stonking great lizards.[4]

However, Komodo dragons belong to a group of lizards called varanids, the only reptiles to cut up the flesh of their victims using their teeth. (Turtles cut up their prey with scissorlike scales surrounding their mouths, but all other reptiles gulp down their prey whole, torn and/or mangled by a bit of perfunctory chewing.) This messy and characteristically mammalian type of butchery has only served to enhance the fearsome reputation of the Komodo branch of the varanid family. A dragon will typically start to dismember a carcass by ripping out the intestines and expelling the contents by shaking them violently from side to side (the least of two good reasons for standing well back when they are feeding). The diaphragm, heart, and lungs are then pulled out of the chest cavity and devoured. Once all the delicacies have been consumed they use their sawlike teeth to cut up the rest of the carcass. At least, this is what they do if the victim is a large animal like a buffalo. Walter Auffenberg, a wholly reliable reporter of dragon ecology and behavior, once saw a 2.5 meter (8.2-foot) dragon swallow a 15-kg (33-pound) eviscerated boar in one gulp. Another 42-kg (93-pound) female nearly doubled her weight in just seventeen minutes by devouring all of a 30-kg (66-pound) boar. Komodo dragons may not be the monsters of legend, but they are clearly extremely powerful and voracious animals.

---

4. A Komodo dragon's weight, however, depends to a large extent on whether, and how much, it has just eaten (see text). In addition, large reptiles of most sorts held in captivity with easy access to lots of food tend to grow faster than they would in the wild. The heaviest accurately measured dragon was a consistently well-fed captive male presented to an American zoologist by the Sultan of Bima in 1928. In 1937 it was briefly put on display in the Saint Louis Zoological Gardens, Missouri, at which point it had attained a length of 3.1 meters (10.4 feet) and a weight of 166 kg (366 pounds).

Dragons will eat any naturally deceased carrion they come across, but they are also highly effective ambush predators, typically lying in wait in dense undergrowth beside forest game trails or concealed by tall savanna grasses. When a deer or wild boar happens past, the dragon bolts from its hiding place to slash at the unfortunate animal with its steak-knife teeth. Often the victim is overwhelmed and killed outright, but sometimes it manages to escape the dragon's clutches and bolt into the undergrowth. However, escapees often suffer raking wounds to their legs and flanks, partial disembowelment, hamstringing (severed tendons in the legs), and, not surprisingly, shock, so they become much more vulnerable to follow-up attacks. Dragons also have particularly poor dental hygiene, so the wounds inflicted on their prey often fester and stink. Once an animal's wounds become infected, dragons simply follow their noses until they find their victim weakened, collapsed, or dead. Infections from animal bites are not uncommon—those from lynx are often fatal to caribou calves, for example—but the slash-and-follow method of killing is highly effective for a large reptilian predator-scavenger with a powerful supercharger but a limited capacity for chasing fast-moving or powerful prey.

Deer are the most common victims of dragons these days, although they also take snakes, birds, rats, dogs, wild boars, goats, horses, water buffaloes, and the occasional unwary human. Wounding, hamstringing, and passing on a mouthful of bacteria are particularly effective against the larger of these animals, because big creatures are more likely to escape an initial ambush. The ability of Komodo dragons to kill large animals in stages like this gives a possible insight into one of the most intriguing aspects of their evolutionary history. The puzzle is that deer, boars, and buffaloes, the major prey items of adult dragons today, were all introduced to Komodo and its neighboring islands only a few thousand years ago by humans. The islands' remaining fauna of predominantly small birds and reptiles hardly constitutes appropriate fare for such a large and ferocious predator. So what did dragons eat before humans spiced up the menu? Fossils from the island of Flores suggest that the only animals that could possibly have sustained populations of Pleistocene dragons were elephants! Granted, they were pygmy elephants, but *Stegodon sompoensis,* the smaller of the two species, weighed as much as a buffalo. Dragons would have scavenged carcasses and actively preyed on juveniles, but adults were probably vulnerable too. A mouthful of flesh may have been a dragon's immediate reward for an attack on a fully grown elephant, but the big prize would come a few days later as the drifting aroma of putrefying flesh betrayed the faltering footsteps of an impending corpse and guided the dragons to an easy feast.

A diet of elephants in the Pleistocene may well have encouraged gigantism in the lizards of Komodo, but there is a more fundamental puzzle about the unusual size of these predatory lizards. On the plains of Africa, animals the size of pygmy elephants, buffaloes, wild boars, and rusa deer are preyed upon by lions, leopards, cheetahs, hyenas, and various types of hunting dogs. The lands in close proximity to Komodo have their complements of large furry predators, including leopards, sun bears, and tigers on the larger Indonesian islands, and big cats, hyenas, and wolves in southeast Asia. But on Komodo there are no large mammalian predators, and as far as we can tell from the fossil record, there never have been. Is this the ultimate reason for the evolution of the fearsome Komodo dragon? Did these monitor lizards evolve into large top predators only because they were nurtured through the early years of their evolution in an environment free of animals with high-powered metabolic engines? We will never know for sure, but the biogeographic and paleontological evidence is at least consistent with this idea. Suspicious, isn't it, that by far the largest species of predatory lizard in the world is stuck out on a few islands in the Indian Ocean with no history of occupation by large mammalian predators? Given these peculiar geographic circumstances, the dragons of Komodo cannot challenge the rule that warm-blooded animals naturally have the edge over cold-blooded ones as large land animals.

What about tortoises? Giant tortoises are the heaviest of all extant reptilian landlubbers, attaining weights of 250 kg (550 pounds) or more. Like Komodo dragons, giant tortoises today are found only on islands without any large mammalian predators, namely the Galápagos Islands off the coast of Ecuador, and half a world away on the island of Aldabra in the Indian Ocean. Could the lack of large mammals in these far-flung places explain why tortoises managed to evolve into large land animals? Probably not, although the isolation of the Galápagos Islands and Aldabra has afforded some protection from mammalian attention. The truth is that there have been many species of big tortoise in the postdinosaurian era, and some of them roamed continental areas in full view of the mammalian hordes. (The first unquestionably tortoiselike fossils appear in the Eocene period, although there is some evidence that the group may have been in existence by the end of the Cretaceous. Around two hundred species of tortoise have been identified in all, of which around thirty are still extant.) Fossil remains of a 2-million-year-old behemoth called *Geochelone (Megalochelys) sivalensis*[5] are widely distributed around the world in fully continental settings. Estimates of the weight of these giants vary widely (wildly, in fact): the most inflated is over 4 U.S. tons, but 1 or 2 tonnes is probably more realistic. Whatever their weight, at 2 meters (6.6 feet) long and 1 meter (3.3 feet) high, these walking Volkswagen Beetles would have dwarfed any living terrestrial reptile. *G. sivalensis* is an extreme case, but many large tortoises, including the South American ancestor of the

5. Sometimes called *Colossochelys atlas* or *Geochelone atlas.*

surviving Galápagos populations, have, at one time or another, lived side by side with large mammals. In fact, tortoises are the only type of reptile since the demise of the dinosaurs to have consistently churned out big terrestrial species in defiance of the prevailing mammalian hegemony.

The success of tortoises boils down to a couple of key adaptations, the most obvious being armor. No other tetrapod in history has taken the defense method of being hermetically sealed to quite the same lengths. Until the arrival of humans, the basic tortoise strategy of being no threat to anyone and far too much trouble to eat was sufficient to frustrate all but the most determined players on the ecological stage.[6] Tortoises are also unusual among terrestrial reptiles in being predominantly vegetarian. Needless to say, with their very heavy armor, tortoises would not make good pursuit predators. Ambush tactics work for sleek and speedy Komodo dragons, but a giant carnivorous tortoise preying on deer would undoubtedly do a lot more waiting than hitting. Tortoises don't need to be sleek, swift, aggressive, or heavily weaponed for the purpose of feeding themselves, so they can pile on protective armor with impunity.

The only other large terrestrial reptiles currently on Earth are snakes, all of the family *Boidae*. Anacondas, three species of python—reticulated, Indian, and African—and boa constrictors may all exceed 5 meters (16 feet) in length, and anacondas and reticulated pythons may occasionally reach 9 meters (29 feet). Thick-

---

6. Golden eagles and bearded vultures in the Middle East regularly kill spur-thighed tortoises by dropping them onto the largest available rock from an altitude of 30 meters (100 feet) or more. The rate of predation can be quite high, especially during the birds' breeding season (a single pair of eagles dispatched eighty-four tortoises by this method over a four-month period). Some mammal predators may have (and may have had in the past) jaws powerful enough to crack open small tortoises, but, as far as we can tell, none has ever made a living out of the practice. The lives of freshwater turtles are much less secure as many have to contend with crocodilians, an omnipresent hazard of tropical freshwater life to which we will turn shortly.

set anacondas may top 200 kg (440 pounds) in weight, which clearly puts them into the megafaunal category by any reasonable yardstick. All these snakes are committed meat-eaters, but unlike Komodo dragons they are widely distributed across tropical and subtropical regions of the Americas, Africa, and Asia in a variety of habitats. How does our rule stand up to the test of large terrestrial snakes? Pretty well, although the arguments are a bit more subtle than for tortoises and dragons.

Snakes in general tend to be more vulnerable to predation than heavily shielded animals like tortoises. Adults frequently meet violent deaths at the hands and mouths of big fish, frogs, predatory lizards, other snakes, army ants, mammals, and birds, and hatchlings run the gauntlet of predatory spiders and scorpions. So populations of snakes survive by suffering a certain amount of wastage. Most snakes are predators of living animals, however, and most will use their offensive weaponry for defense in a clinch. All snakes have mouths studded with teeth, and many species also load them with poison. The coiled body of a rattlesnake, the elevated and expanded head of a cobra, or the red, black, and yellow stripes of a Texas coral snake, are such obvious displays of potential danger that all but the most specialized predators can usually be discouraged.

But for all their poisons and posturing, the most common form of defense among snakes is to be inconspicuous. It is their unique body shape and cryptic patterning and coloration that allow these animals to secrete themselves in the environment, await or sneak up on their prey, and avoid the unwelcome attention of others. As we saw in Chapter 2, few warm-blooded animals have gone down the evolutionary path toward being long and skinny, for reasons of energy conservation. But snakes steal the majority of their body heat from the environment, which means that they can run their metabolic engines on tiny amounts of fuel, be as thin as they like, and conceal themselves much more effectively than bulkier animals. The big five boids are just as cryptic and secretive as most nonpoisonous snakes, despite being protected from predation to some extent simply by being large and immensely powerful,[7] but boids also use their unique bodies to garner resources in habitats that are very different from the hunting grounds of similar-sized predatory mammals with which they might otherwise compete.

Anacondas from South America and reticulated pythons from southeast Asia and Indonesia, for example, are the typical jungle snakes, relying on the dense cover of their tropical rain forest homes to lie in wait for peccaries, capybaras, deer, antelopes, wild boars, cattle, and sometimes people. Once fed, these slow-moving animals retreat even further into the protective embrace of the forest, sometimes for weeks on end, to digest their victims. Both species are semiaquatic, and anacondas in particular are frequently found in slow-moving rivers, vegetation-choked lakes, and swamps, floating motionless in the water waiting for the unwary. Boa constrictors, despite the inaccurate way that they are portrayed in bad adventure movies, are the smallest of the big five boids and completely harmless to humans. They are also denizens of tropical forests throughout Central and South America, where their range overlaps with that of anacondas. Constrictors are more widespread than their larger cousins and exploit a wider range of habitats. Too slow to chase their prey, constrictors rely on their camouflage to lie in wait for passersby. African pythons are mainly denizens of the tropical forests of west Africa, and are equally at home in the trees, in dense undergrowth, and in freshwater, the last-named habit earning them their alternative names of rock or water pythons. They are occasionally found on savanna grasslands, where their ground-hugging form keeps them concealed. Indian pythons

---

7. Hungry jaguars and caimans may occasionally attack anacondas, but the outcome of such battles is anything but a foregone conclusion—anacondas have been known to constrict and kill caimans up to 2 meters (6.6 feet) in length.

prefer wooded areas ranging from dense evergreen forest to more open deciduous woodland. They are excellent climbers, often ascending trees to seek out and ambush prey. Like their larger Asian and South American cousins, they are also common around rivers, lakes, and marshes. Indian pythons eat mainly small mammals, but given the opportunity they are capable of taking on much larger animals.

The branches and canopies of trees, dense undergrowth, leaf litter on the forest floor, swampy ground, and vegetation-choked water—these are the typical haunts of large snakes. Most large mammals are just too bulky and hungry to exploit such difficult habitats, where the only alternative to sit-and-wait predation is to crash through the undergrowth or canopy or splash through the water in search of victims. Certainly, large predatory snakes and mammals may interact from time to time, and they may even eat each other if the opportunity arises, but they nevertheless manage to coexist primarily by the simple expedient of keeping out of each other's ecological way. In fact, it is difficult to imagine two groups of 30-kg-plus (66-pound-plus) meat-eating animals living in the same geographic region and taking similar-sized prey, yet managing to avoid each other so completely. Being 7 meters (23 feet) long but only 15 cm (6 inches) tall is a uniquely reptilian way of making a living, and one from which warm-blooded predators are always likely to be excluded. Our rule stands up to the challenge of snakes.

Komodo dragons, giant tortoises, and boids are the only dry-land reptilian giants currently on Earth, but there is one part of the world with such an unusually high diversity of reptiles, including a number of moderately large species, that it deserves some comment. The unique reptile fauna of the arid lands of Australia has been the source of much puzzlement among ecologists over the years. Just among the larger varanids currently roaming the region are spotted tree goannas, mangrove monitors, sand goannas, water monitors, lace monitors, and perenties ranging in length from 85 cm to 2.4 meters (33 inches to 8 feet). There are around seventeen species of monitor lizards alone in Australia, far more than in climatically similar regions in Africa and North America, and many of the medium-to-large ones seem to fill niches in Australia that elsewhere are occupied by animals like weasels, ferrets, mongooses, badgers, dogs, and cats. The lack of large mammalian predators in Australia is particularly striking and puzzling: the only native carnivores weighing 5 kg (11 pounds) or more around today in the region are the Tasmanian devil, a general scavenger, and the spotted-tailed quoll (like a small cat). In contrast, east Africa has twenty mammalian carnivores above 5 kg in weight, North America twenty-seven, and the relatively tiny region of Thailand twenty-four.

Perhaps not surprisingly, given these statistics, biologists of previous generations thought that there must be something in the makeup of the marsupial branch of the mammalian sisterhood—perhaps their relatively small brains—that has prevented them from evolving into large carnivores like their placental cousins elsewhere.[8] But South America once supported all manner of preda-

---

8. Australia is rightly famous for its marsupials, but, in fact, 50 percent of the continent's native fauna consists of unpouched placental mammals, half of which are bats and the other half rodents. The recent discovery of a fossil bat from Murgon in southeast Queensland suggests that these consummate mammalian travelers have been in Australia for most of the Cenozoic (at least 55 million years), while rodents have probably been scampering around the continent since the late Miocene. There is also controversial evidence from Murgon that condylarths—a mammalian group that on other continents evolved into animals like whales, cows, horses, and dogs—might also have been present in Australia early in the Cenozoic. It is often assumed that the prevalence of marsupials in Australia is a reflection of how difficult it has been for putatively more "advanced" placental mammals to colonize this isolated island continent. The paleontological evidence challenges this view, however, and some researchers now suspect that marsupials came to dominate Australia because they outcompeted their early-Cenozoic placental cousins.

tory marsupials, including the ecological equivalents of animals like bears, dogs, wolves, civets, and saber-toothed cats, and Australia itself supported a rich variety of large marsupial carnivores earlier in the Cenozoic when much of the continent was covered with complex and floristically diverse tropical rain forests. In other words, the marsupials-can't-hack-it idea is clearly untenable. If there is no intrinsic barrier to the evolution of large marsupial carnivores, why are there so few in Australia? Find an answer to this question, and the prevalence of large reptiles in the region may begin to make more sense.

In 1981, A. V. Milewski put forward the ingenious idea that Australia's unusual tetrapod communities may ultimately be explained by the aridity and unpredictability of the Australian climate, combined with the nutrient-poor nature of the continent's soils.[9] In the semi-arid parts of North America, for example, the concentration of phosphorus in soil is usually above 300 parts per million (ppm), and none has concentrations lower than 100 ppm. But in Australia vast tracts of land are extremely poor in phosphorus, with levels sometimes as low as 25 ppm. Phosphorus is one of the essential elements required for plant growth, so low concentrations in the soil can limit the productivity of whole ecosystems. The reasons for the infertility of Australian soils are firstly that the granites, gneisses, and sandstones of Australia are naturally rather low in phosphorus, so any soil derived from them will necessarily be phosphorus-poor too. Secondly, Australia has had an unusually stable tectonic history, with few significant mountain-building events in recent geologic time, so there aren't many areas of uplifted rock exposed to weathering and erosion. In the absence of fresh inputs of chemical nutrients from the ero-

---

9. For an alternative interpretation, although not necessarily contradictory, see Wroe (1999).

sion of high ground, the slow process of leaching by percolating rainwater and groundwater over the eons has made the already deficient soils of Australia even less fertile.

How might infertility favor predatory reptiles over mammals? The answer seems to revolve around the intermediate effect of infertile soils on plants, particularly in arid regions. Plants grow slowly where there is a lack of chemical nutrients, and they often evolve tough, resinous, prickly foliage. Manufacturing hard, chemically unpalatable leaves is more "expensive" than making soft, flimsy ones, but it has the advantage of discouraging attacks by grazers. This allows plants to hang on to most of the precious nutrients that they manage to strain from the wretched soil and gives them a better chance of getting through their slow life cycles unmolested. Low plants on infertile soils also tend to be tough perennials rather than weedy annuals, because the nutrients needed for rapid seasonal growth simply aren't available. So, surprising as it may seem, there tends to be *more* vegetation permanently stationed on arid, infertile soils than on arid, fertile ones, and this is certainly true of Australia compared with similar climatic regions in North America and southern Africa.

The type of foliage produced by these plants tends to be highly flammable and usually suffers catastrophic burns every ten to twenty years. The speed with which plants re-establish themselves after such events is unpredictable because regeneration depends to a large extent on the availability of water, and rainfall in the arid regions of Australia is rather unpredictable in itself. Add in the influence of the El Niño cycle, which can cause profound and prolonged periods of drought, and the Australian outback becomes one of the most climatically unpredictable places on Earth. The upshot of all this is that Australia's vegetation is particularly unproductive, particularly unpalatable to herbivores, and disappears altogether over large areas at frequent but irregular intervals.

Mammals in general, with their high-powered and fuel-hungry

metabolic engines, depend on a large and constant supply of food whether they be herbivores or carnivores. If the environment is unproductive, the herbivores will be less common and more widely scattered, and as fertility and plant productivity decline further, a point may be reached where the density of herbivores is so low that a viable population of carnivores cannot be supported. The effects of unpredictable disturbances just exacerbate the situation, because if fire destroys 90 percent of an already impoverished grassland, then only a few herbivores will be able to survive on what remains. A few may be enough for the herbivore population eventually to recover, but it is unlikely that a viable population of predators could survive on such a doubly depleted food resource. So areas characterized by low productivity *and* severe resource bottlenecks are likely to be bad for mammals in general, and disastrous for predatory mammals in particular.

But cold-blooded consumers, with their lower food requirements and ability to fast for long periods, should be at an advantage under such conditions. This seems to be borne out when the faunas of areas with similar climates but different levels of soil fertility are compared. Seed-eating rodents, for example, are eight times more abundant in dry regions of North America and southern Africa than in climatically alike parts of Australia. Similarly, larks, small bustards, and pheasants are absent from the arid parts of Australia but common all year round in similar parts of Africa. The lack of small warm-blooded plant- and seed-eaters in Australia may account for the scarcity of animals that specialize in catching these animals, like predatory mammals, birds, and certain types of snakes. The main small-bodied foliage- and seed-eaters in arid regions of Australia are not rodents and birds but cold-blooded invertebrates—leaf-eating insects, for example, consume a greater proportion of the foliage growing in Mediterranean (semi-arid) Australia than in similar parts of southern Africa. Australia also has an enormous variety of ants, many of which gather seeds. The most common type of predator of these invertebrates is the same as in most other arid areas of the world: lizards. But in Australia, the range of opportunities for lizards is particularly broad because of the variety of invertebrate prey, the lack of predatory mammals and birds, and the hiding places available within and beneath the foliage of slow-growing, persistent, harsh-leafed plants. Australia has more slow-moving, nocturnal species of lizard than anywhere else in the world, probably because of the lack of predation by, and competition with, nocturnal mammals. Australia has more ant-eating lizards than anywhere else, and many sit out in the open and wait for ants to blunder into their path, a strategy that would undoubtedly invite predation in areas with high densities of carnivorous mammals and birds. The dense and persistent ground cover also provides homes and hunting grounds for numerous small, delicate lizards such as skinks, and these, in turn, support predatory lizards. At the top of the food chain the larger lizards of arid Australia thrive because they mostly eat other lizards, which are probably neither common nor reliable enough as a food resource to support monitor-sized mammals.

So if Milewski is right, the reptile-rich fauna of Australia may be due to the advantages of cold-bloodedness under conditions of aridity, climatic unpredictability, and low soil fertility. If so, then we might expect the faunas of nutrient-poor areas under the same climatic conditions to have evolved in similar ways regardless of where in the world they are. Milewski realized that this would be an important test of his argument, so he compared and contrasted the faunas of study areas in western Australia and South Africa with similar climatic regimes and very low soil fertility. He discovered some striking similarities. Mammals are scarce in both areas, and burrowing rodents, large hoofed animals, and predators of large animals are all absent. The common daytime lizards of the two regions are similar both in appearance and habits, and the snakes appear to be ecologically similar too, even though the South

African ones are vipers and the Australian ones the distantly related elapids. Of course, Milewski found many differences between the faunas of the two regions, but he was able to relate most of these to rather obvious differences in flora, food resources, and predators. For example, shrubs producing seeds attractive to ants are more common in Australia and, not surprisingly, ant-eating lizards are more prevalent too. Plants with underground storage organs such as bulbs and tubers are more common in South Africa and support populations of mole-rats, a type of animal with no Australian analogue. South African reptiles are typically more heavily armored and the snakes more aggressive than their counterparts in Australia, which probably reflects the greater diversity and abundance of mammalian predators such as mongooses, cats, foxes, and baboons.

Milewski's fascinating theory is workable, clearly testable in its details, and intriguing in its implications about the nature of life on our planet. If he is right, then the implication is clear: cold-blooded metabolic engines, even when fitted to reasonably large animals in open terrestrial environments, can sometimes be superior to warm-blooded ones. The reason that mammals are so dominant as large animals almost everywhere else may simply be that almost everywhere else is more fertile and climatically stable than modern Australia.

And there is nothing essentially unusual about Australia. The infertility of this chunk of continental crust is really just an accident of history. In the future, Africa could just as easily break free of its moorings and drift off into isolation. If it avoided collisions with other landmasses so that mountains never formed, the topography would eventually erode away (or gradually "flow" back toward flatness, as modern geologic opinion would have it), nutrients would be leached from the soil, and the average fertility of the land would probably decline. If these processes continued uninterrupted for tens of millions of years, what would become of the elephants, giraffes, antelopes, lions, and hyenas that currently dominate the grasslands of this continent? Would they gradually give way to giant tortoises and monitor lizards with metabolic engines better suited to the poverty and unpredictability of the environment? It is an intriguing thought. What if all the continental landmasses came together to form one huge supercontinent as they did at the end of the Permian period (Fig. 3.1)? Such a pileup would undoubtedly produce mountain ranges, but it might also prevent any significant continental drift for many millions of years afterward. If the system remained jammed up for long enough, large areas might change inexorably toward the sort of infertility that characterizes Australia today. The interiors of large landmasses also tend to turn into deserts because clouds coming in from the sea drop their rain before reaching very far inland, so the Earth could end up with a supercontinent depleted in nutrients *and* arid over much of its surface. What would happen then? Could the world then experience a real Age of Reptiles, a time when large cold-blooded land animals turn the tables on their warm-blooded cousins? For those who cling to ideas of progress in evolution—the amphibian-reptile-mammal-human sequence of alleged increasing superiority—such an idea can only seem ludicrous, but put aside such notions, as surely we should 140 years on from Darwin, and such an evolutionary trajectory is clearly within the bounds of possibility. Natural selection just picks horses for courses, and there are certain courses where cold-blooded metabolic engines outdo warm-blooded ones. Who knows what the future holds for our ever-changing planet and the creatures chosen by the underlying logic of natural selection to grace its surface?

In our search for an explanation for why there are so few large reptiles wandering the land areas of our planet, we have thus far concentrated just on living animals and communities. But the modern world, of course, is just a shaving from the tip of geologic history. What about the past? Is the fossil record consistent with the idea that large reptiles cannot compete with large mammals in most terrestrial situations? The record does, in fact, tell a similar story, but there are a few intriguing reptiles that may qualify as exceptions.

The first, a giant lizard, survived almost down to the present day in that paradise for large reptiles, Australia. As its name suggests, *Megalania* was not your typical wall-walking or hole-hiding lizard. The most recent estimate for the weight of these reptilian predators is around 1 tonne. Apart from their size, they seem to have resembled modern Komodo dragons in many important respects (Fig. 4.7), being typically lizard-shaped with long tails, sprawled limbs, and massive jaws lined with serrated teeth. Given the damage that Komodo dragons inflict on deer and buffaloes with their relatively puny oral weapons, the attack of a 1-tonne *Megalania* must have been a uniquely terrifying, if mercifully brief, experience. What did *Megalania* eat? As the ecologist David Quammen once half-jokingly remarked, probably anything it wanted. Giant kangaroos and wombats seem to have been the most likely quarry. Was it an ambush predator? We have no way of knowing, but if *Megalania* had a typically reptilian metabolic engine and preyed on kangaroos, then it is difficult to see how it could have done anything else. Some paleontologists believe that *Megalania* may have been a semiaquatic predator of the water's edge, the equivalent of modern-day adult Nile crocodiles. Alternatively, it may have been able to rush kangaroos and other mammalian prey from hiding places on dry land just like modern-day Komodo dragons. The fossil remains of *Megalania* are frustratingly rare and fragmentary, so we have no choice but to await further finds before attempting to interpret the ecology of this fascinating creature in any greater detail.

The other reptilian sore thumbs that paleontologists have chipped out of the fossil record must be the most curious post-dinosaurian reptiles of all. Pristichampsine crocodiles flourished from the late Cretaceous to the Eocene (Fig. 3.1) in many parts of the world, and a superficially similar, but rather distantly related, line of crocodile-like animals called sebecosuchians were common in South America from the late Cretaceous to the Pliocene. Both groups had a suite of very uncrocodile-like anatomic features. Fossil and modern crocodilians usually have conical teeth for grasping and immobilizing their prey, and they dismember their larger victims either by thrashing them around or by whirling around underwater using the resistance of the water against the carcass for leverage. But pristichampsines and sebecosuchians, just like predatory dinosaurs and Komodo dragons, had laterally compressed teeth with serrated back edges, which suggests that they sliced up carcasses on land. Common vertebrate fossils found in association with early pristichampsines hint that they preyed on mammals, including the early relatives of horses. Were they freshwater ambush predators like modern crocodilians? Possibly, but they had unusually long legs for crocodilians, tails that were round in cross section, and hooflike claws, all of which suggests that they did a lot of moving around on land. Terrestrial crocodiles capable of chasing and killing small horses? No wonder these reptiles are commonly described by paleontologists as "enigmatic," a standard scientific euphemism for problematic.

These animals appear to have reached their zenith just after the dinosaurs had become extinct but before the later Cenozoic plethora of large mammalian carnivores had evolved. There were obviously vacancies for large terrestrial predators at this time, and as crocodilians had come through the Cretaceous mass extinction in better shape than most, they were ideally placed to take up the challenge. However, once large mammalian predators had evolved, the pristichampsines of Europe and North America seem to have disappeared rather quickly. No one knows whether the two events were causally linked, but the timing is suspicious. It may also be significant that the one place where large terrestrial crocodiles managed to survive almost down to the present day was Australia. Teeth and skull fragments of a crocodile given the name *Quinkana* have been found in Australian rocks of Pliocene to late-Pleistocene age and, who knows, may even have preyed on early human settlers. What a uniquely scary place Australia must have been for mammals in the Pleistocene, with both *Quinkana* and *Megalania* lurking—well, wherever it was that these puzzling animals habitually lurked.

Komodo dragons, giant tortoises, boid snakes, and the curious fauna of Australia are all qualifiable exceptions to the rule that warm-bloods tend to have the edge over cold-bloods as large animals in open terrestrial environments. As for *Megalania* and the various types of extinct terrestrial crocodile, we simply know too little about them to be sure how they lived. Given the time at which they came to prominence and/or their survival only in places which, for various reasons, seem to be particularly well suited to animals with low resource requirements, they probably also rank as qualifiable exceptions even on the basis of our fragmentary knowledge.

When challenged by the evidence of extant and fossil animals, the pattern of warm-blooded domination of most terrestrial ecosystems as large body sizes holds well enough to be formalized, with appropriate caution, into a general rule of life. Big animals restricted to living out their lives on the two-dimensional plane of the ground find it difficult to avoid the attention of others. Even in forests where the opportunities for concealment are greatest, hiding is invariably easier for small animals than large ones. As body size increases, animals gravitate toward the ground and the lines of sight for both predators and prey gradually lengthen. With no holes big enough to hide in, no obstacles big enough to crouch behind, and no third dimension to exploit, wait-and-hit becomes a less viable method of predation and hiding in or bolting to inaccessible places becomes a less viable method of escape. If large predators want food, they generally have to work hard for it, and if large plant-eaters want to avoid becoming food, they have to work even harder. As lines of sight lengthen, the survival tactics of fleeing, chasing, and fighting become progressively more important, and warm-blooded animals are simply the consummate chasers and fighters of the animal world.

But you and I and the rest of the mammalian sisterhood are lucky that the world is mostly a rich, fertile, and relatively stable place. Our metabolic engines keep us warm enough to react to any threat or opportunity at a moment's notice and deliver enormous amounts of sustained power when we need it. Yet the cost of our ecological abilities is a cruel feedback loop of hunger, fed by activity, fed by hunger, hour after hour, day after day, for our entire lives. Like the Red Queen in *Through the Looking Glass,* we must run and run just to remain in the same place. Most terrestrial environments at the present time provide enough food on a reliable basis to favor our frenetic approach to life, but it didn't have to be this way and it needn't be so in the future. If the climate warms, as it has in the past, if soil fertility declines and food supplies on the major continental landmasses become disrupted by irregular disturbances, then our tenure as the dominant large land animals on Earth could be under threat. And from whom? From the progeny of our sprawling reptilian ancestors whose physical endurance and athletic abilities seem so pitiful, but whose physiological endurance makes us look pitiful. On a constantly changing Earth, and under the blind and unprogressive rule of natural selection, no future is certain.

But at present the rules of the ecological game for large animals over most of the Earth's dry surface gives us the edge over our reptilian cousins. As large land animals we can be proud of our achievements and revel in our ecological talents. But we should not forget that the land is just one component of the terrestrial environment. Without the water provided by rain and collected by streams, rivers, and lakes we would all perish in very short order. It is to these freshwater ecosystems that we turn next to discover that the rules of life for large animals on land have no relevance whatsoever.

[7]

THE ARTERIES OF THE LAND

The inland waterways, lakes, and swamps of the Americas are home to alligators, four species of crocodile, and six species of caiman.[1] The freshwaters of Africa are patrolled by dwarf, slender-snouted, and Nile crocodiles (Fig. 7.1), those of the Indian subcontinent by mugger crocodiles and gharials, and those of the Indo-Pacific by false gharials, Siamese, Philippine, New Guinea, Johnston's, and estuarine crocodiles. Crocodilians are the best-known reptilian giants currently lurking in the freshwater ecosystems of the world, but they are by no means the only ones. New World waterways support alligator snapping turtles weighing up to 80 kg (175 pounds), Arrau River turtles, common snappers, Central American river turtles, Florida soft-shells, peninsula cooters, and a host of physically less imposing species. The turtles of Africa, India, and the Indo-Pacific include huge emydids, Nile soft-shells, southern Asian soft-shells up to 1.2 meters (3.9 feet) in length, and a particularly unpredictable species of narrow-headed

---

1. American, Orinoco, Cuban, and Morelet's crocodiles; common, yacare, broad-snouted, black, Cuvier's dwarf, and Schneider's dwarf caimans. The family *Crocodylidae* is divided into three subfamilies. The *Alligatorinae* houses the American and Chinese alligators and six species of caiman. The *Crocodylinae* contains twelve species of "true" crocodile, one dwarf crocodile, and the false gharial. The *Gavialinae* is home to just one species, the Ganges gharial.

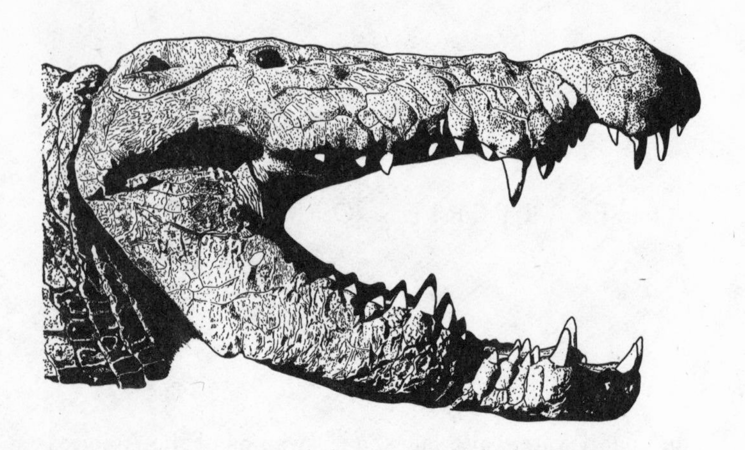

FIGURE 7.1 The imposing gape of the mighty Nile crocodile. Length up to 5 meters (16 feet).

soft-shell that has been known to attack goats. Add to the list pirarucu, Nile perch, tarpon, long-nosed gar, giant tigerfish, pike, zander, and a host of other huge piscine hunters, and the freshwater ecosystems of the world seem quite overrun with large cold-blooded predators.

But where are all the mammals? The only fearsome species found in freshwater today is the common hippo, a creature with a foul temper and the body weight to back it up. Angry hippos have been known to decapitate Nile crocodiles by rolling them around in their mouths, but they cannot be considered competitors of crocodilians in any meaningful sense because they forage on land and eat grass. In fact, a survey of all the world's freshwater ecosystems reveals not a single surefire mammalian contender for anything approaching a crocodilian way of life. Duck-billed platypuses, water opossums, various shrews, and hare-lipped bats acquire some or all of their food from freshwater, but all are small creatures that specialize in taking small items of prey. Beavers, muskrats, water rats, coypus, and capybaras spend much of their

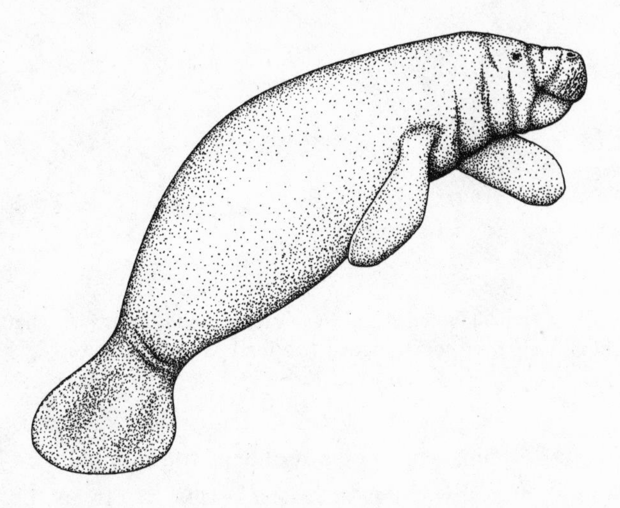

FIGURE 7.2 Ox manatee. Length up to 3 meters (9.8 feet).

time in and around water, but they also forage on land and are predominantly or entirely vegetarian. Jaguars, tigers, a few other cats, jaguarundis, raccoons, raccoon dogs, and some bears may hunt in the shallows, but all are really terrestrial predators with an opportunistic outlook when it comes to food. Otters, mink, and Congo water civets have more aquatic lifestyles, but all are relatively diminutive,[2] take predominantly small items from the water, and also hunt on land. The only mammals completely committed to life in lakes and rivers and rivaling crocodilians in size are ox manatees from South America (Fig. 7.2) and four species of freshwater dolphins (Fig. 7.3). Manatees, like hippos, are dedicated vegetarians, so they are not even potential competitors of crocodilians, but dolphins—70- to 160-kg (155- to 350-pound) freshwater predators—just might be. We shall find out in due course.

2. Male giant otters from South America are the heaviest freshwater mustelids at 25 to 30 kg (55 to 66 pounds).

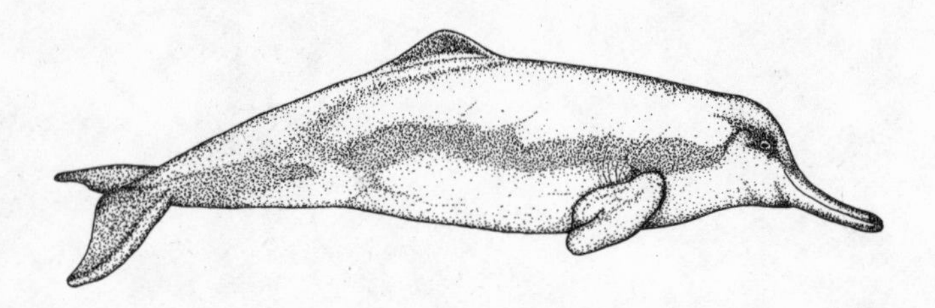

FIGURE 7.3 The baiji, a species of freshwater dolphin from the Yangtze River, China. Length up to 2.5 meters (8.2 feet), weight up to 160 kg (350 pounds).

Leaving dolphins aside for a moment, the dominance of large cold-bloods in freshwater is at least as conspicuous as the domination of dry land by large mammals. So striking is the difference that one might reasonably expect scientists to have figured out long ago why the creatures in these two great terrestrial environments are so different in their engine specifications. I assumed as much in the 1970s when I first became enamored of crocodilians, but twenty years of fruitless searching have established that the issue has received hardly any attention at all. Why this should be so is puzzling, but it probably reflects the fact that professional ecologists tend to specialize either in land or freshwater ecology and thereby manage to avoid the broader scientific questions straddling the divide. It may also have something to do with the fact that large freshwater animals of all kinds are strongly concentrated in tropical and subtropical regions,[3] whereas the majority of sci-

3. Low-latitude freshwater ecosystems form the focus of this chapter because these are the ones that have been available for exploitation by animals without significant interruption since the end of the Cretaceous period 65 million years ago. Contenders for the niche of large freshwater predator at low latitudes have had the longest time to gain ascendancy in head-to-head competition. In contrast, land farther north than the Great Lakes in North America and Germany in Europe spent much of recent geologic history under thick layers of ice.

entists over the last few hundred years have lived and worked in temperate Eurasia and America. I hasten to add that a definitive explanation of why large cold-bloods outshine their warm-blooded cousins in rivers and lakes will not be forthcoming here either, but I hope that what follows will at least elucidate the nature of a fascinating unsolved problem.

The most obvious place to start an investigation of this kind is with the different physical properties of water and air and the consequences of immersing different types of animals in each. Since water sucks heat out of warm objects very quickly, a hot animal spending most or all of its time in a river or lake would be faced with only four workable options: abandon thermoregulation and live with a body temperature close to ambient; occupy only the warmest waters to minimize the rate of heat loss; manufacture and maintain an extremely effective layer of insulation; or increase metabolic heat production.

As far as we know, the first option has not been adopted by any aquatic mammal. Manatees seem to exploit a combination of the second and third: they have unusually low metabolic rates for mammals, but they compensate with big bodies, fatty insulation, and by seeking out water above 18°C (64°F) (below 15°C [59°F]) manatees quickly become hypothermic and die. Fast-moving, predatory otters, in contrast, exploit the third and fourth options: they use air-filled, oil-impregnated fur rather than fat for insulation, and they also have substantially higher metabolic rates than most land-living mammals of similar size. Interestingly, the actual rates seem to reflect the amount of time spent in the water—around 50 percent higher than expected in river otters, and two and a half times higher in sea otters.

Despite such elevated rates of heat production, it is an otter's insulating pelt that appears to be the crucial adaptation. Until quite recently, around 35 percent of otters held captive in zoos died prematurely of pneumonia and associated infections because no one realized that they needed substantial areas of land in their enclosures for drying and maintaining their fur. One attempted relocation of sea otters ended in disaster because the animals were tightly crated up on the journey and prevented from grooming. By the time they were released, their pelts had deteriorated so badly that they quickly became soaked to the skin. Some of the unfortunate animals were recaptured, but the others succumbed to hypothermia and died. To prevent the same thing happening in the wild, otters spend enormous amounts of time keeping their fur in peak condition. Sea otters spend around 10 percent of their lives performing all sorts of tricks and contortions to ensure that their pelts stay fully waterproof. They twist and turn to groom and aerate the fur on their backs, roll through rafts of air bubbles generated by vigorously beating the water, squeeze water out with their hands, and even manually inflate their fur by blowing into it. Being a sleek, furry, warm-blooded predator in cold water involves an enormous amount of routine maintenance work just to stave off death from hypothermia.

An aquatic existence in all but the warmest water clearly presents some tricky problems for mammals. In contrast, the physical properties of water seem particularly well suited to cold-bloods. Water has a much greater thermal inertia than air, which means that it changes temperature more slowly and thus tends to remain within a much narrower temperature range, so aquatic reptiles can avoid the extreme and potentially life-threatening thermal highs and lows with which their relatives have to contend on land.[4] And by heating their bodies behaviorally, and allowing themselves to cool into a semitorpid state when there is no other option, reptiles also avoid

4. Although most crocodilians keep themselves warm by the simple expedient of occupying tropical and subtropical environments, some are able to cope with remarkable drops in temperature without expending any energy in thermoregulatory countermeasures. The geographic range of American alligators, for example, extends northward into the temperate zone where winter ice often forms on the surface of still

the high metabolic price paid by aquatic warm-bloods like otters.

So just by considering some of the basic thermal consequences of immersing warm-blooded and cold-blooded animals in water, we already begin to sense a shift in the balance of power. On land big warm-bloods seem to have all the advantages: we can move our heavy bodies around terrestrial environments at speeds that would exhaust cold-blooded animals of similar size within moments, and by staying at a constant temperature we can remain active at times of the day, night, or season when the athletic abilities of cold-bloods would be even further compromised. But in water the advantages seem to be whittled away from both ends: reptiles benefit because water provides thermal stability, allowing them to operate at or near peak efficiency for a greater proportion of the time, while aquatic mammals incur extra energetic costs over and above the already profligate expenditure of their land-living relatives. Being large, manufacturing fat, and being thermally unadventurous in terms of habitat choice is the relatively low-cost solution employed by manatees, which works well if the water is stocked with food that doesn't swim away. But for a river otter, whose fate is largely determined by the efficiency with which it can overtake blisteringly fast animals like fish, being fat and slow is obviously a less realistic option. A substantial increase in metabolic rate coupled with a thermal shield of oil-impregnated, inflatable fur requiring constant care and attention seems to do the trick, but you do begin to wonder whether otters might not be

---

waters. Radiotelemetric studies have shown that alligators faced with extreme cold seek out quiet backwaters and position themselves on shallow banks with their nostrils poking above the surface. By using their snouts as snorkels they keep a breathing hole open even when the surface water freezes solid. Some individuals have even been found with their snouts frozen into the ice. This doesn't seem to be a problem provided their airways remain open; they just lie around and wait for the thaw to set them free. Implanted temperature sensors have registered core body temperatures down to 5°C (41°F) during the coldest spells, and the animals recover with no apparent ill effects. Alligators are as tough as old boots.

better off chasing rats on land. Could the answer revolve around something as simple as this? Is it the hostile nature of water to animals with warm bodies that explains why big mammals do so poorly in the freshwater ecosystems of the world?

No. And we only have to follow the course of a river to its destination to see why. The oceans of the world are just as wet and often much colder than lakes and rivers, yet they are home to all manner of large mammals. The global tally of 113 species consists of 78 cetaceans (whales, dolphins, and porpoises), 33 pinipeds (seals, sea lions, fur seals, and the walrus), the sea otter, and a marine version of the manatee called a dugong. Conversely, if water is such an ideal medium for reptiles, the oceans of the world should be teeming with them. Not so. Although the species count at 80 is superficially quite impressive, 71 of these are sea snakes, of which 70 are restricted to just one area of ocean between India and northern Australia (the pelagic sea snake also inhabits this region but ranges across the Pacific and westward to the east coast of Africa). The 9 remaining marine reptiles consist of 7 turtles, a semiaquatic iguana from the Galápagos Islands, and the estuarine crocodile from the Indo-Pacific. Large mammals eclipse large reptiles in diversity, abundance, and biomass both on land and in the sea, which leads us right back to square one in terms of figuring out why the reverse is true in freshwater.

We need to take a different tack. What do land and sea have in common that freshwater lacks which could explain these curious biogeographic patterns? The most straightforward possibility involves the deep evolutionary history of reptiles and mammals. In the Mesozoic the most numerous large tetrapods everywhere were reptiles: crocodilians in freshwater,[5] plesiosaurs, icthyosaurs, and

---

5. A few strange-looking dinosaurs with very long, narrow snouts and crocodile-like teeth called spinosaurids are also known from Africa, Brazil, and Europe. The most recently discovered, dubbed *Suchomimus tenerensis,* was an animal 11 meters (36 feet) long, 2.5 meters (8.2 feet) at the hip, and generally tyrannosaur-like or birdlike in

FIGURE 7.4 From top to bottom: ichthyosaur, plesiosaur, and pliosaur

mosasaurs at sea (Fig. 7.4), and dinosaurs on land. By the end of the Mesozoic, dinosaurs and the largest marine reptiles had all gone extinct, but freshwater crocodilians came through into the dawn of the Cenozoic with relatively minor losses.[6]

---

form. That *Suchomimus* was an obligate fish-eater is an idea that I find hard to swallow, but because of the shape of the jaws and teeth this has to be the default interpretation.

6. A number of theories have been advanced to account for this differential survivorship, all of them rather speculative. Perhaps the strongest relates to differences in the nature of food chains. Recall that the Mesozoic Era probably ended with the impact of an extraterrestrial body that threw millions of tons of dust into the atmosphere, darkening the Earth and shutting down photosynthesis. This disaster would have hit terrestrial and marine ecosystems particularly hard because green plants form the base of the food chain in both cases. But the functioning of watercourses is usually heavily

To the ragbag of small mammals that crawled out of the forests at the end of the Cretaceous, establishing an evolutionary beachhead on land and in the sea was largely a matter of seizing the moment, but Cenozoic freshwaters must have been a different prospect altogether: for mouse-sized insectivores to evolve into fearsome carnivores in ecosystems free of significant competition is one thing, but for them to do the same in an environment with a crocodile around every meander is surely quite another.

Could this be the answer? Are large mammalian predators scarce in freshwater simply because crocodiles got there first? It is an intriguing theory. The problem with it from a scientific standpoint is that it appears to be untestable. What sort of evidence could possibly reveal the crocodiles-got-there-first idea to be false? The lack of large freshwater mammals at latitudes beyond the physiological tolerances of modern crocodilians might suggest that mammals struggle with freshwaters in general, but then inland waters at high latitudes are likely to differ from tropical ones in age, glacial history, water temperature, flow regime, productivity, chemistry, and so many other factors that this line of thought always fizzles out rather quickly. The scarcity of freshwater mammals

---

reliant not on "green" food chains but on "brown" ones, sequences of eater and eaten underpinned by dead organic material either produced in-stream or washed in from the land. Primary consumers fuel their lives using this material, they are eaten by secondary consumers, and so on up the food chain to animals like otters. The extent to which brown food chains can keep freshwater ecosystems functioning without inputs from planktonic plants, marginal vegetation, and microflora is very poorly understood, but we do know that such chains contribute a much greater proportion to the total productivity of watercourses compared with either the land or the sea. Under the impact scenario, the Earth would have been plunged into darkness for months or years, causing green food chains to collapse. The supply of detritus and dissolved organic matter into watercourses, however, would probably have continued long after land plants and the creatures dependent on them had perished. Life in rivers, and the lakes and swamps fed by them, would undoubtedly have been shocked and transformed by the impact and the far-reaching environmental changes that followed in its wake, but brown food chains *may* have held the whole system together long enough for the dust to clear and for some semblance of ecological normality to return.

could just have different explanations in different parts of the world. No other potential tests of the theory spring readily to mind. Of course, just because the crocodiles-got-there-first idea is difficult to test does not mean that it is wrong: the natural world, after all, cares not one jot about the philosophy of science. My own feeling, for what it is worth, is that the survival of predatory megareptiles in freshwaters at the end of the Cretaceous and their extinction almost everywhere else is one of the main reasons for the current lopsided distribution of large cold-bloods and warm-bloods between land, sea, and freshwater. Of course, being untestable, this is a very safe point of view.

Clearly, however, we need to explore some alternatives before accepting this rather weak explanation. For the sake of argument, therefore, let us assume that differential survivorship across the Mesozoic-Cenozoic boundary cannot be the whole answer. Is there any ecological mechanism that might explain why big reptiles do so well in freshwater?

Perhaps. Recall that mammals require much more food than similar-sized reptiles and succumb very quickly if the supply of food is interrupted for any reason. Recall also how A. V. Milewski used these two basic facts to construct a plausible theory to account for the unusually reptile-rich fauna of Australia. With reasons to believe that animals with low-powered metabolic engines might do particularly well in relatively unproductive or unstable environments, it makes sense to explore the nature of terrestrial, marine, and freshwater ecosystems in these terms before giving in to the crocodiles-got-there-first idea.

All the animals with which we are concerned in this chapter rely directly or indirectly on plants for sustenance, and the amount of this crucial fuel churned out by different types of ecosystems varies enormously. Open tropical oceans, for example, produce fifteen to twenty times less plant material per unit area per year than tropical rain forests, and the reason is not immediately obvious; both environments are brightly lit by the sun and in general have access to sufficient amounts of all the essential chemicals required for plant growth. The difference in the amount of vegetation present at any one time—called the "standing crop"—boils down to differences in environmental stability. Because the sea is so restless, plants are constantly under threat of being thrown up on the land, dragged into freezing polar regions, or pulled down so deep in the water that there is insufficient light for photosynthesis. There isn't time for plants to live out long lives in the ocean, so they adopt the ecological strategy of rushing through their life cycles at breakneck speed in the attempt to produce another generation of offspring before disaster strikes.[7]

Because oceanic plants rapidly grow, reproduce, die, and sink, the amount visible in the water at any one time is always small. In contrast, the land provides firm, reliable foundations, so plants can slowly grow to enormous sizes, manufacture showy flowers and

---

7. The life-history strategies of oceanic and rain forest plants represent the extremes of a continuum reflecting the degree to which environments suffer disturbance. Arable fields, for example, are among the most unstable of terrestrial environments because they are periodically plowed up. Owing to this seasonal trauma, the only plants that manage to gain a foothold, other than the ones planted by the farmer, are various types of annual "weed." Weeds grow just tall enough to poke their heads above the soil before channeling all their remaining energy into producing as many seeds as possible. Once the seeds have been abandoned to the wind, the parent plant dies and trusts to luck that a few of its offspring will land on patches of bare earth to start the cycle again. In contrast, fully formed oak trees never suddenly spring up in the middle of arable fields for the simple reason that they need decades to get through their life cycles. The odd acorn may find its way into a cornfield and begin to grow in the newly plowed earth, but some piece of farm machinery will inevitably squash it long before it can set seed. But in more stable environments, oak trees outcompete weeds because their ecological strategy is based on growing vigorously to shade out the opposition before settling down to produce a steady stream of acorns over hundreds of years. This is why much of the land at temperate latitudes before the arrival of agricultural humans was covered with oak trees and not weeds. It is a general ecological rule that the more disturbed the environment, the weedier and more hurried the dominant plants tend to be. Arable fields bristle with the most frenetic of terrestrial plants, but nothing on land can compare to the constant trauma suffered by the plants of our restless oceans.

succulent fruits, and indulge in all manner of time-consuming activities that would be impossible in a constantly churning ocean. The exposed heartwood of an old rain forest tree and the youngest leaf in its canopy together represent the accumulation of plant material not over a few hours or days but over many centuries.

But if the land is home to large plants with a slow rate of turnover, and the sunlit layer of the sea to small plants with a high rate of turnover, why doesn't the amount of plant tissue produced equal out over the course of a year? The answer to this question lies in the fate of the nutrients used by plants to build their bodies. Rain forest leaves end up on the forest floor, where they are broken down by worms, other small invertebrates, fungi, and bacteria, and the nutrients released in the process are quickly vacuumed up again to make new leaves, twigs, roots, and sapwood. A rain forest soil may be extremely poor in nutrients at any given time, but this is largely because it is constantly being sucked into a state of infertility by a veritable and literal jungle of plants. Oceanic plants also scavenge nutrients from their immediate surroundings, but on death their corpses sink to the inky depths, usually within the bodies of grazing animals.[8] So the sunlight zone of the sea remains poor in nutrients not because they are constantly being recycled into new plant production but because they are continually being lost to the oozes and muds of the sea floor. The slow churning of the ocean system does eventually bring a few nutrients back up into the light, where they can be reused by subsequent generations of plants, but this happens mainly in a few small areas of the ocean system, such as the shallow banks of the North Sea and the areas

---

8. Typtically 80 percent or more of the delicate nutritious planktonic plants of the oceans are consumed by herbivores—mainly microcrustaceans but also sedentary filter-feeders and large vertebrate filter-feeders like whales—a percentage three times higher than in grasslands and fifteen times greater than in forests. Most of the tough, fibrous, cellulose-ridden, often poisonous plant material produced by terrestrial ecosystems is bypassed by grazers and enters the soil, where it is broken down by decomposers.

of upwelling off the coast of Peru and Antarctica, where the geometry of the ocean basins and movement of currents encourage nutrient-rich water to rise to the surface. Even in these places lack of particular nutrients and/or light, combined with the violence of the ocean swell, often limits plant productivity to levels well below those of the most productive environments on land.

Despite these differences in standing crop and annual productivity, however, the land and the sea are similar in one very important respect: the supply of plant material is pretty reliable in space and time. The region of ascending, nutrient-laden water around Antarctica, for example, is always rich in plant and invertebrate life in the summer when the sun shines on it, which is why many seals and whales migrate there every year in anticipation. And while the present position of the continental landmasses and ocean basins remains roughly the same, this area of the sea will always provide. Marine mammals may have to travel enormous distances to access the richest sources of food, but then they live in a highly connected ocean system that presents few barriers to their wanderings and in a gravityless medium that takes much less energy to move through than air.

Similarly, tropical rain forests on land churn out between 1 and 3.5 kg (2.2 and 7.7 pounds) of plant tissue per square meter (1.2 square yards) annually, temperate grasslands between 0.5 and 1.5 kg (1.1 and 3.3 pounds), and production stays more or less constant at these levels from one year to the next. True, the land is susceptible to droughts, floods, fires, storms, and the constant cycling of the seasons, but populations of large, mobile animals can usually find somewhere within running or walking distance to survive even such severe disturbances. The savannas of Africa, for instance, are subject to pronounced annual dry seasons when the grasses die back and food becomes scarce, but populations of large herbivores manage to survive by switching to more resilient vegetation nearby or by following the path of the rains across the continent.

The amount of vegetation produced by different types of freshwater environments is very variable. Plant production in standing water is usually low for much the same reasons that it is low in the oceans, and animals at the top of the food chain tend to be scarce as a consequence (as any angler with a passion for large trout will attest). Swamps, in contrast, rank among the most productive habitats, with some tropical examples churning out plant material as fast as rain forests, at least in years when water levels are favorable. The productivity of rivers and streams is thought to be low compared with most dry-land habitats, but, in truth, we understand less about the ecology of watercourses than about almost any other type of environment on Earth. Careful comparisons of the annual productivity of rivers and their adjacent flood plains would be of great interest to ecologists interested in patterns of animal life within and between these environments, but such studies simply do not exist. In fact, there is little agreement among researchers even on what constitutes plant production in watercourses, let alone on how to measure it. A global average of 0.25 kg (.55 pound) of plant tissue per square meter (1.2 square yards) per year for rivers and lakes combined is widely quoted, but the estimates for running waters are usually based just on the production of algae attached to objects on and in the stream bed. Vegetation growing along the shallow margins of watercourses may be highly productive, but, then again, apart from a few species of fish such as *Tilapia* and grass carp, virtually nothing seems to eat it. Why this should be so, when similar plants on land suffer enormous grazing pressure, is another of the enduring mysteries of ecology. Marginal vegetation cannot simply be dismissed, however, because a proportion of it ends up in the food chain as detritus or dissolved organic matter and supports a rich community of insects, crustaceans, fungi, and bacteria. Much of the algae in watercourses ends up as detritus too, along with variable amounts of dead plant tissue washed or blown in from the land. Once again, our understanding of the inputs, losses, processing, and assimilation of this material is still in its infancy. The general perception of running waters as considerably less productive than adjacent habitats on land is probably true in the majority of cases, but clearly a lot more research needs to be done.

Although our understanding of the productivity of watercourses is frustratingly poor, it is clear that these environments differ from marine and terrestrial ones in two important respects: they are isolated from each other by the topography of the land and the presence of the sea, and they are often highly unstable. Watercourses are notoriously unpredictable environments, where erosion and abrasion, siltation and burial, flood, drought, and extremes of water quality are ever-present dangers. Seasonal rainfall cycles affect all the habitats in a drainage basin, but the effect on rivers is usually the most dramatic. The Belize River in Central America, for example, is subject to a four month dry season—moderate by tropical standards. Its discharge may vary from a torrential 200 cubic meters (262 cubic yards) per second in September and October to a dribbly 5 to 20 cubic meters (6.5 to 26 cubic yards) per second in May. Over the course of centuries or millennia there are bound to be periods of severe drought when large stretches of the Belize dry up altogether. In fact, in most small drainage basins and many larger ones in various climates, zero flows in rivers occur quite regularly, so virtual or complete disappearance of the primary habitat on which freshwater animals rely is a hazard against which they must have some sort of defense.

And wet seasons may be no less traumatic. The last time the Brahmaputra flooded, two-thirds of Bangladesh ended up under water. On the Amazon and Orinoco in South America, watercourses regularly break up into chains of shallow lagoons at the height of the dry season, but flooding in wet months can submerge the surrounding forests to depths of up to 9 meters (29 feet) for many kilometers (miles) from the main channels. Superimposed on these seasonal cycles are unpredictable periods of storm flow when streamside trees, marginal vegetation, and attached algae may be ripped up and dumped in the ocean, large sections of the riverbed scoured away and redeposited, and huge amounts of sediment lifted into suspension, plunging the stream bed into darkness and shutting down photosynthesis for days or weeks. All in all, the arteries of the land are profoundly difficult places in which to eke out a living.

What characteristics might allow populations of animals routinely to survive disturbances of the severity regularly visited upon freshwater environments? One important factor promoting the long-term survival of many species is the ability to move to safe refuges in times of crisis. The temporary disappearance of watercourses and wetlands in a region would undoubtedly have a severe impact on resident and migrant bird populations, but the unique mobility of these animals at least allows them to search the landscape effectively for some safe haven. Otters too are highly mobile and surprisingly adaptable to different environments and food resources. Although traditionally thought of as eaters of fish and aquatic invertebrates, European and American otters have also been known to turn their paws to frogs, toads, salamanders, newts, crayfish, ducks, coots, moorhens, waders, starlings, swallows, water voles, shrews, moles, rats, mice, mink, muskrats, beavers, and, in Scotland and southern England at least, large numbers of rabbits. Otters also move long distances over land and frequently exploited coasts before human development drove most of them away. Coastal environments may not only provide refuges in troubled times but also a route from one river system to another. In a similar vein, both west African and American manatees inhabit shallow seas as well as the lower reaches of rivers and estuaries, so they may also be able to escape to coastal refuges and migrate to less disturbed areas when the environment turns against them.

Many crocodilians and turtles are also known to move overland when rivers disappear during the dry season, and most species of croc have been spotted in brackish and saltwaters at one time or another, so there are a number of possible escape routes open to these creatures too. Nile crocodiles, for example, are occasionally encountered in the sea, and they have also been known to walk up to 10 km (6.2 miles) each night in search of water, hiding in thick undergrowth during the day to escape the heat and the attention of lions. But unlike any freshwater mammal, Nile crocodiles can, and frequently do, dig themselves into the mud of dried-up riverbeds if no water is available and simply wait for the rains to return. They may stay entombed like this for months under normal circumstances, often accompanied in their burrows by turtles adopting the same passive strategy, but large adult crocs could probably survive holed up like this for a couple of years. Mugger crocodiles of the Indian subcontinent may find refuge in brackish and coastal marshes, and they also wander extensively on land during the dry season seeking out the last remaining pools of water, but if they fail to find any, they too retreat to burrows, sometimes 8 to 10 meters (26 to 33 feet) below the parched surface, and simply wait for the next cloudburst.

This remarkable physiological endurance is surely the key factor that has always given tetrapod cold-bloods that crucial edge in the global freshwater system. The ability of individual crocodilians to survive long periods of enforced starvation is the most dramatic illustration of their physiological capabilities, but it is probably at the population level that the most important benefits of cold-bloodedness accrue. Because reptiles can survive on small amounts of food and fast through regular lean periods, they should be able to colonize and maintain adequate breeding populations in freshwater environments that are far too poor and unpredictable for warm-bloods. In turn, this should allow them to spread over wider areas, occupy a broader range of habitats, and be more abundant within the limits of their physiological tolerances than high-powered animals of similar size and diet.

The distribution of large reptiles and mammals in the world's freshwater is at least consistent with this idea. Crocodilians and turtles are still very widespread in tropical and subtropical waterways of all kinds, and crocodilians in particular were much more widely distributed a few hundred years ago before humans began decimating the populations of most species. In contrast, ox manatees are denizens of blackwater and oxbow lakes and lagoons regularly and reliably flooded by the mighty watercourses of the Amazon, and river dolphins are found only in the most persistent reaches of massive continental river systems, namely the Orinoco and Amazon in South America, the Ganges and Indus in India, and the Yangtze in China.[9] Several species of marine dolphins and porpoises enter estuaries and the lower reaches of rivers from time to time, which is how the world's current handful of freshwater species first evolved, but only in the largest, most permanent ones have they managed to make the complete transition to freshwater life and hang on to the present day. Clearly, if dolphins and manatees are exceptions to the global pattern of freshwater domination by large cold-bloods, then there are good reasons for believing that they are qualifiable exceptions.

The ability to maintain large populations under a broad range

---

9. The finless porpoise probably has the least restricted geographic range of all cetaceans in freshwater environments, being found in river systems and lakes (if connected to rivers) throughout the Indo-Pacific from Japan to northern Australia. Moreover, these animals are frequently observed in watercourses known to be ephemeral over timescales of tens to hundreds of years. But unlike ox manatees and dolphins, finless porpoises also roam the oceans between river systems, so their extensive geographic range is not the result of adaptation to the rigors and vicissitudes of freshwater life but of opportunistic exploitation of freshwater environments wherever and whenever they happen to be favorable.

of environmental conditions over a wide geographic area also acts as a very effective buffer against the sorts of profound environmental changes that might otherwise drive a species toward extinction. Wide-ranging populations composed of many individuals are invariably more difficult to eradicate than small, geographically restricted ones, so the risk of localized adversity turning into total extermination is likely to be much reduced. A comparison of the recent fortunes of crocodilians and dolphins in the face of human persecution provides a particularly telling example. Over the last few hundred years, and particularly in the last fifty, crocodilians have been the target of human abuse on a massive scale. Most species have been hunted ruthlessly either for "sport," their skins, or to protect people and livestock from predation. To make matters worse, all crocodilians have suffered alteration and destruction of their habitats by expanding human populations hungry for space, natural resources, and somewhere to flush their industrial and domestic pollutants. The American alligator, for example, once abounded in all but the smallest streams and isolated ponds across Texas, Louisiana, Arkansas, Mississippi, Alabama, Georgia, Florida, North and South Carolina, and as far north as Virginia, but hide-hunting, persecution, habitat destruction, and pollution took a heavy toll on their populations, eliminating them from many areas and bringing the species close to extinction in most others. Similarly, the geographic range of the Chinese alligator has shrunk by 90 percent just in the last twenty-five years as a result of pollution, dam construction, conversion of wetlands to agriculture, hunting for hides and traditional medicines, and human intolerance in general. Hunting of black caimans in the Amazon basin for skins and for the dubious purpose of protecting cattle has pushed this species firmly onto the endangered list too. Orinoco, Philippine, and Siamese crocodiles are now classed by conservation authorities as critically endangered, gharials and Cuban crocodiles as endangered, and American, mugger, and dwarf crocodiles as vulnerable. Even aided by appropriately emotive terminology, the sheer scale and rate of the slaughter in recent decades is difficult to grasp.

But the remarkable and uplifting aspect of this otherwise miserable story is how some crocodilians have not only survived the onslaught of the human race but even managed to stage spectacular comebacks in some parts of the world. Small surviving populations of American alligators, for example, mostly in areas away from centers of human population and beyond the reach or interest of developers, have seeded a recovery that has seen this species regain much of its former range. By some estimates there are now more than 380,000 alligators in Louisiana alone, and a hunting season has had to be reinstated to protect valuable fur-bearing animals such as otters and muskrats. Estuarine crocodiles in New Guinea and Australia too have shown a similar pattern of massive decline followed by recovery at unanticipated rates in response to reduced levels of persecution. Unfortunately, the prospects for Chinese alligators are much less certain, but small populations have managed to survive, like their American counterparts, by retreating to refuges in the lower part of the Yangtze River that have remained, at least until very recently, beyond the commercial interest of humans because of severe annual flooding.

All in all, there is considerable hope these days that with further legislation and continuing conservation, many of the world's crocodilians may be saved—or rather save themselves—from global extinction. Such cautious optimism is a far cry from the hopelessly bleak situation that in 1971 prompted Wilfred Neill to write:

> I doubt that any crocodilian species will persist in nature beyond the present century. Even if there is an abrupt surge of interest in the activities of crocodilians, these survivors from the Age of Reptiles are likely to vanish before their biology is even half understood. The next reviewer of crocodilians will probably find

> himself relying almost wholly on earlier publications for any information that cannot be derived from museum skins and skulls.

One year from the new century at the time of writing, I am happy to report that all the crocodilian species over which Neill lamented are still with us. I do not wish to convey complacency or advocate anything except the strengthening and extension of legislation, but it has to be said that crocodilians have proved a lot more difficult to kill off than anyone in the 1970s dared hope. Conservationists have played an important part, particularly in richer countries like the United States, but it is probably the ability of crocodilians to survive in virtually any old stream or puddle if left alone that has allowed them to sidestep, for the moment at least, Wilfred Neill's apocalyptic prediction.

In contrast, humans have never had any particular genocidal urges toward any of the world's inoffensive river dolphins, yet the restriction of these animals to particular types of habitat, and thus to relatively small parts of the global freshwater system, has amplified the effects of routine human encroachment and precipitated disastrous declines. The three populations of boto—also known as the Amazon River dolphin, pink porpoise (erroneously), or pink dolphin—from the Amazon and Orinoco basins and upper Madeira River are probably least at risk at the present time, but their medium- to long-term survival is still in the balance as hydroelectric development, deforestation, overfishing, and pollution of watercourses continues apace in Venezuela, Colombia, Ecuador, Bolivia, Peru, Guyana, and Brazil. Hunting of Ganges and Indus dolphins for "dolphin oil" used as a fishing lure has been implicated in the declines of these species, but other environmental changes brought about by humans have probably been much more damaging, and both species are now endangered. The Yangtze dolphin appears to be on the verge of extinction, and the Mississippi dolphin has already passed unobtrusively into the fossil record. There are simply no out-of-the-way refuges for animals capable of surviving only in the most stable reaches of the largest continental river systems.

The resilience of crocodilian populations in the face of intense human persecution is without doubt just an indication of the qualities that have allowed these remarkable creatures not only to flourish in some of the most disturbed and unpredictable environments on the planet, but also to survive extraterrestrial impacts, refrigeration of the Earth during pulses of glaciation, sea-level change, tectonic upheavals, profound changes in climate, and 200 million years of competition from several dynasties of otherwise highly successful terrestrial animals. Moreover, it is *because* of the nature of their metabolic engines, not despite them, that crocodilians have proved themselves virtually indestructible. This interpretation also happens to fit neatly within the conceptual framework constructed by A. V. Milewski to account for the reptile-rich fauna of Australia (see Chapter 6): if factors as subtle as low plant production, periodic disruption of food supplies by fire, and delayed regeneration because of unpredictable rainfall are sufficient to shift the balance of power in favor of large reptiles on land, it seems entirely plausible that the profound disturbance regimes to which the majority of freshwater systems are subject might result in a similar pattern of dominance. Indeed, if anything, Milewski's theory seems more obviously applicable to freshwaters than to the arid terrestrial environments that formed the focus of his original studies.

After 200 million years of hegemonic ecological success, it is time to recognize crocodilians for what they truly are: perhaps the closest approximation to an unsurpassable ecological design in the entire history of tetrapod life. Warm-blooded animals have the capacity to evolve into highly successful aquatic animals, as the current global tally of marine mammals attests, yet in 65 million years warm-bloods have not managed even to loosen the crocodilian supremacy over tropical and subtropical freshwaters. The fundamental superiority of crocodilians within this particular ecological sphere—a superiority rooted firmly in the nature of their metabolic engines—is as true today as it was when their forefathers were pulling dinosaurs into the rivers and swamps of the Mesozoic. And given just a little elbow room by the most intelligent, powerful, and indiscriminately destructive species ever to walk the planet, they will undoubtedly be among the front-runners to witness the dawn of the Telozoic too—whenever that happens to be.

[8]

TAKING WING

The Cenozoic Era (Fig. 3.1), that stretch of geologic time bounded by the Cretaceous asteroid impact and the present day, is traditionally called the Age of Mammals. The justification for this is the spectacular diversification of mammals over the last 65 million years and the evolution of numerous species large enough to be highly conspicuous both as living animals and as fossils. The rise of our closest relatives in the Cenozoic has certainly been impressive, but we should not get carried away. For example, from a recently published encyclopedia of animal life we learn that "there are about 4000 species of mammal . . . the most adaptable and diverse group of vertebrates on our planet today."[1] The most adaptable? Perhaps. But adaptability is a slippery concept and rather difficult to quantify. It would be a close-run thing with birds, whatever the measure. And the most diverse? Nowhere near. The Earth currently supports around 4325 species of amphibians, 6900 reptiles, 9700 birds, and 45,000 fish. Mammals are the *least* diverse group of vertebrates on our planet today. If diversity be the yardstick, and if we must limit the discussion to backboned animals alone, then we are clearly in the Age of Fish, just as we have been for nearly all of vertebrate history.

---

1. Whitfield (ed.). *Illustrated Encyclopedia of Animals* (1998).

Mammals are not the only group of vertebrates to have diversified markedly during the Cenozoic, either. With a pattern of proliferation even more explosive than that of mammals, and resulting in more than double the number of extant species, a strong case could be made for calling the Cenozoic the Age of Birds. And if birds really did evolve from theropods, as now seems virtually certain, then the hierarchical logic of biological classification implies that we have never really left the Age of Dinosaurs. The Mesozoic monsters may have shrunk a bit and sprouted wings, but they are still the most diverse group of tetrapods on the planet.

So neither present-day diversity nor rates of Cenozoic proliferation can justify the "Age of Mammals" tag. The only other possibilities are mammal-centrism among the chroniclers of evolutionary history, and the fact that mammals have been highly successful at producing large, conspicuous species. Mammal-centrism is a pervasive and obvious bias, but the fact that mammals have churned out big species more often than birds is intriguing. Extant megamammals include both surviving elephants, many primates, and numerous species of perissodactyls (horses, tapirs, and rhinos), carnivores (dogs, cats, hyenas, bears) and artiodactyls (pigs, peccaries, hippos, camels, deer, giraffes, cattle, antelopes, sheep, and goats). Yet of the 9700 species of birds currently on Earth, only the ostrich, the emu, and three species of cassowaries pass the 50-kg (110-pound) mark (Fig. 8.1). A few larger birds have existed in the recent past, and we will meet some of them presently, but, as a group, Cenozoic dinosaurs have clearly come off second best to mammals at being big.

Yet birds are driven by powerful metabolic engines, maintain their bodies at a high and constant temperature, and have extremely high aerobic capacities, so they are closer to mammals in basic energetic design than any other group of tetrapods. In fact birds are rather warmer than mammals internally (39°C to 42°C [102°F to 108°F]), have relatively larger hearts, and a through-flow

FIGURE 8.1 (a) Ostrich; (b) emu; (c) cassowary. All may reach weights in excess of 50 kg (110 pounds).

type of respiratory system that makes the dead-end, bellows-like lungs of mammals look distinctly unsophisticated. In other words, the capacity of birds to sustain activity with oxygen-based metabolism is at least as high as, and often higher than, that of mammals, so the failure of birds to take their fair share of the Earth's megafaunal roles cannot be explained by the nature of their metabolic engines.

Nor should we forget that being a small warm-blood entails substantial energetic expenditure. The metabolic engines of small birds tick over much faster than those of big ones, so they must consume a greater proportion of their body weight in food every day. True, big animals have to eat more food in total just because they have more cells to keep supplied with fuel, but then they also have greater mobility, longer reaches, and bigger mouths with which to do so. And big animals profit from size-related benefits such as better temperature stability, the option of growing thicker insulation, larger stores of food and water, longer fasting times, and decreased vulnerability to predators at very high body weights. Of course, the benefits to a warm-blooded animal of being big do not always outweigh the costs—there would not be so many small warm-bloods around otherwise—but the potential pluses are so obvious that it is no surprise to find that thousands of mammal species over the last 65 million years have found themselves heading down the evolutionary path toward gigantism. But why so few birds? Warm-blooded animals of all kinds are subject to the same laws of mathematics and physics, so why have birds so rarely traveled the same evolutionary road?[2]

One of the reasons can be observed in the rather obvious tal-

---

2. The emphasis of the following discussion is on the evolution of large terrestrial species (ca. 20 kg [44 pounds] or more) and why mammals have produced so many more than birds. For an intriguing theory about how the modal body sizes of birds and mammals may be related to the rigors of flight and the energy available for reproduction, see Maurer (1998) and references therein.

ents and limitations of living birds. A sparrow, for example, jumps into the air and flies off with little more than a flick of its wings. The clattering noise accompanying the takeoff of a pigeon, however, is caused by such vigorous flapping that the animal's wrists bang together behind its back. Vultures cannot jump-fly at all, so they hop forward into the prevailing wind with their enormous wings outstretched. Swans are among the most magisterial of birds when gliding across the water or flying overhead, but the way in which they get from one state of grace to the other by careering across a lake can hardly be called elegant. Extrapolating from the weights and takeoff speeds of animals that can fly, an ostrich would probably need to run at 1000 km (620 miles) per hour just to get off the ground.

The reason that taking to the air is more of an effort for big birds than small ones relates, once again, to the scaling relationship between areas and weights. In a series of birds of increasing size but similar shape, body weights naturally increase much faster than wing areas, which means that the weight supported by each square centimeter of wing, called the "wing loading," tends to be greater for large birds. In theory, this imbalance could be counteracted by increasing wing area at a greatly accelerated rate as birds evolve toward large size, but in practice this would rapidly produce wings so heavy and unwieldy that they could not be operated at all. The big-wing solution works for man-made gliders, some of which have wingspans of over 25 meters (82 feet), but the design concept is obviously workable only up to a point for animals that have to flap their wings to get into the air. The only alternative to inoperably large wings is to increase air speed to generate more lift. Jumbo jets travel down runways faster than single-seater aircraft, and it is for much the same reason that swans run across lakes, and hawks, eagles, vultures, and many shorebirds jump off trees or cliffs.

And the problems for heavy animals do not end once they are

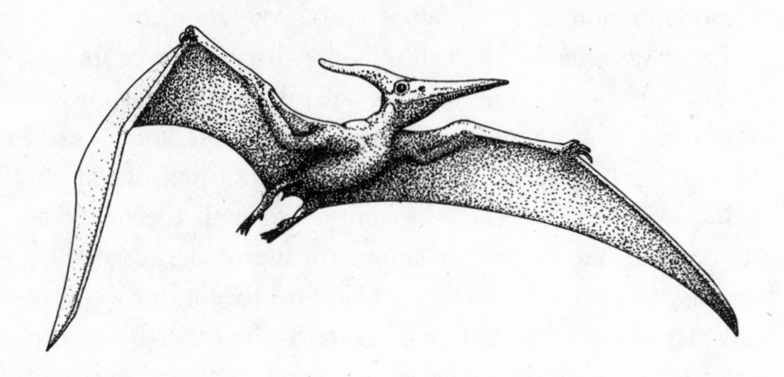

FIGURE 8.2 *Pteranodon*, the largest pterosaur known from reasonably complete skeletons, with a wingspan of 7 meters (23 feet) or more. Its skeleton was so lightly constructed that it probably weighed no more than 15 kg (33 pounds). This is about the weight of a kori bustard, the largest living bird capable of flight.

in the air. Birds of all sizes need to produce roughly the same amount of power to keep each kilogram of themselves aloft. As we saw in Chapter 1, resting metabolic rates decrease with increasing body weight, and an animal has to increase its metabolic rate well above resting levels to sustain flight. The metabolic engine of a 4-gram (.14-ounce) hummingbird, for example, has to tick over three times faster when flying than when at rest, but a 7-kg (15-pound) vulture has to increase its metabolic rate by a factor of twenty. The engines of terrestrial mammals run ten to fifteen times faster than normal during very heavy exercise, so, taking into account the fact that birds have relatively larger hearts and lungs, the biggest birds currently capable of maintaining level flight in still air are probably expending energy at close to the theoretical maximum. The difference in the energy required for flight and that available is also why vultures, storks, and many other aerial leviathans spend much of their time cheating gravity by gliding around in thermal updraughts. The larger flying pterosaurs of the Mesozoic probably exploited the same principle (Fig. 8.2).

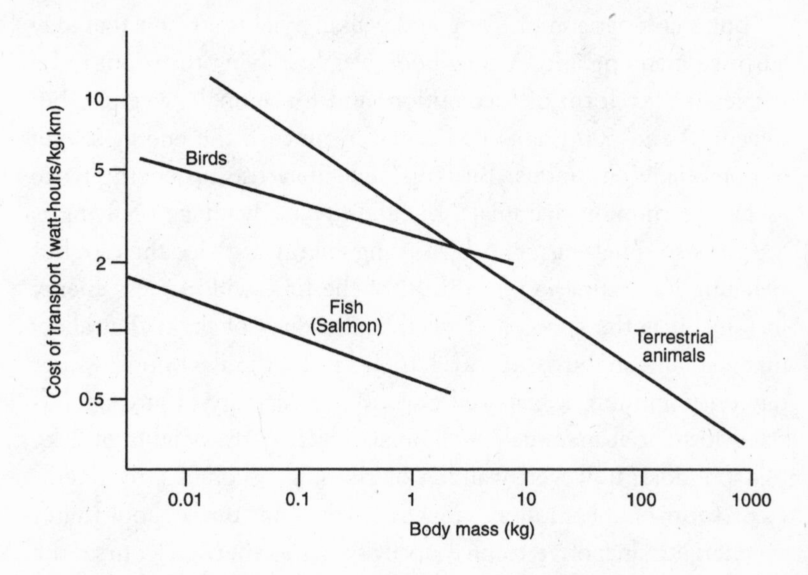

FIGURE 8.3 Redrawn from McMahon and Bonner (1983). (1 kg = 2.2 pounds; 1 km = 0.62 miles).

The energy consumed by a 100-gram (3.5-ounce) bird in flight is around twice that of a similar-sized walking mammal, but this doesn't mean that flying is necessarily an inefficient way of getting from A to B. Figure 8.3 compares the energetic cost of moving around in air, on land, and in water. Numbers on the vertical axis record the minimum amount of energy required to move 1 kg (2.2 pounds) of an animal through a distance of 1 km (.62 mile) at the most energy-efficient speed. The energy-sapping influence of gravity means that both walking and flying are very expensive ways of moving relative to swimming. The extensive transoceanic migrations of many marine mammals appear to be impressive feats of endurance, especially to creatures like us who are poorly designed for moving through water, but for an animal with the anatomic wherewithal, swimming is by far the most energy-efficient way of getting from place to place.

But a comparison of flying and walking yields a result that may surprise many people. At low body weights, flying turns out to be the less costly form of locomotion, and for animals weighing between 10 and 100 grams (.35 and 3.5 ounces), the energy saving is potentially enormous. Birds in flight may use up energy twice as fast as running mammals, but the great advantage of flying is that it is so much faster. A bird using energy at twice the rate but reaching its destination in a third of the time will use less energy in total. It is the speed and relative efficiency of flight that allow many small songbirds to travel 1000 km (620 miles) nonstop per day on migration, a feat well beyond the capacity of any similar-sized terrestrial mammal. As animals reach body weights of 2 kg (4.4 pounds), however, walking or running becomes a more efficient form of locomotion, suggesting that big birds should have an energetic incentive to give up flying altogether. Of course, the benefits to a bird of retaining the power of flight may well outweigh the costs even at body weights much greater than this, because although it may be more energy-efficient for a 7-kg (15-pound) vulture to walk around the savanna rather than fly, getting to a carcass while there is still some meat left on it is obviously the prime objective, and taking to the sky is a very useful ability in this regard. (In reality it would be more efficient for a vulture to walk only if it moved like an animal adapted for terrestrial locomotion. Similarly, although swimming is an efficient form of locomotion for fish and aquatic mammals, it is a very inefficient way for humans to get about because we are the wrong shape, our muscles are in the wrong places, and our extremities make poor paddles.)

The upshot of all this is that small birds buy the power of flight relatively cheaply and gain a wide range of benefits in the process. These include the ability to forage over wide areas, exploit three-dimensional habitats such as forests, escape the attentions of ground-living predators and competitors, migrate to seasonally optimal environments at relatively low energetic cost, and avoid localized environmental catastrophes such as fires, floods, and droughts. Equally clearly, if birds are to evolve into really big animals, then they have no choice but to give up their aerial endeavors along with all the concomitant advantages. No such sacrifice accompanies the evolutionary path toward gigantism in mammals (except for bats), which alone could be enough to explain why so many more of us have evolved into really big species.

This simple explanation for the scarcity of megabirds can be contrasted with a number of others that have been proposed which involve judgments—often dubious—of the relative merits of various mammalian and avian characteristics. For example, the putative success of Cenozoic mammals in general, and of large mammals in particular, has often been attributed to their teeth. The myriad ways in which these hard, enameled structures have been modified to suit particular circumstances has allowed mammals to adapt to almost every type of foodstuff available. Mammalian teeth have evolved to perform different jobs not only between species but also within the mouths of individual animals. Many small mammals have rows of small, sharp teeth ideal for puncturing the tough body casings of insects. Others possess wide, thickly enameled molars for crushing shellfish. Dolphins and porpoises have numerous conical teeth for catching slippery animals like fish and squid. Predatory cats and dogs have large canines for holding and subduing their victims and blade-like cheek-teeth for slicing flesh into manageable chunks. Rodents have two pairs of continuously growing incisors with hard enamel only on the front surface; as the softer dentine behind wears down, the enamel forms a chisel-like cutting edge that can make short work of even the toughest wood and nuts. But perhaps the most important characteristic of mammalian jaws and teeth is that our molars and premolars meet head-on and can move laterally across each other:

in other words, mammals can chew, an ability unique to us among living tetrapods. The explosive proliferation of omnivorous and herbivorous mammals in the Cenozoic, particularly rodents, horses, pigs, camels, deer, cattle, antelopes, sheep, and goats, is often attributed to the seemingly unlimited capacity of mammalian teeth to be modified to process different types of plant material.

In contrast, birds have beaks. Granted, these adaptable structures have become modified into everything from parrotlike nutcrackers to nectar-sipping needles, but many biologists have felt that the adaptive scope of what is essentially a simple horny cone split down the middle is rather limited compared with tooth-lined jaws. In particular, the inability of birds to chew means that any mechanical processing of plant material has to take place internally. Birds use tough, muscular gizzards lined with a hard material called koilin and often filled with pebbles for grinding up food. These gastric mills may be extremely powerful, but, once again, some biologists have found it hard to shake off the suspicion that they do not really match up to the elegant cranial cuisinarts of mammals. This suspicion is strengthened by the fact that bulk processing of low-energy plant material like grass is extremely unusual among birds[3] but very common among mammals. The conclusion seems straightforward: for dedicated herbivory at least, teeth appear to be better tools than gizzards.

For all anyone knows, this conclusion may be right.[4] But there

---

3. Screamers, hoatzins, and owl parrots are primarily leaf-eaters, while some *Galliformes,* ratites, geese, plantcutters, and rails include much green plant material in their diets.

4. Of course, it may be wrong: post hoc rationalizations of this kind should always be treated with caution. There may, in fact, be mechanical and energetic advantages to be had from gizzards in some circumstances. Elephants may spend up to sixteen hours a day eating in order to fuel their enormous bodies, and most of this time is spent chewing. Many sauropod dinosaurs, on the other hand, were much bigger than elephants but had smaller heads and mouths, yet some paleontologists believe that these animals were nevertheless able to process enough plant material to fuel high metabolic rates. For this to be energetically possible, some sort of internal processing over and

is another interpretation which, right or not, at least has the merit of avoiding value judgments. Recall that small warm-blooded animals lose heat from their bodies very easily, and that their metabolic engines have to tick over fast in order to compensate. The smaller an animal gets, the faster its engine races, and the greater the amount of food it must consume relative to its weight. One way in which small animals can ease this burden is to target energy-rich foods such as seeds, fruits, nuts, and the flesh of other animals. This is what nearly all small warm-bloods actually do. A shrew simply would not be able to liberate energy fast enough from the sort of low-quality herbage consumed by cattle to keep its metabolic engine running at the required rate. But as animals increase in size, heat loss becomes progressively less of a problem and the amount of food required to fuel each cubic centimeter of flesh falls. This gives larger animals the option of eating either small amounts of energy-rich food or large amounts of energy-poor food. In other words, total reliance on low-quality plant food is only possible for warm-bloods above a certain size and becomes progressively less risky as a feeding strategy as animals get bigger. The relevance of this to the size of birds is simple: more than any other type of feeding strategy, bulk processing of low-quality plant material is a game for big animals, and big animals, for all their talents, cannot fly. The scarcity of avian grazers, therefore, may tell us little more than we already knew: flight is such a neat trick that we should not expect to encounter flightless birds very often.

All in all, given the potential benefits of flight and the profound modification of the avian body-plan to accommodate it, it is perhaps not surprising that the vast majority of the world's birds weigh less than 1 kg (2.2 pounds). It certainly makes no sense to interpret the preponderance of large mammals on Earth as an in-

above that employed by elephants would appear to be required, and gizzards would do the job nicely. See Bakker (1986) for an extended discussion.

FIGURE 8.4 Skeleton of *Diatryma steini* from the lower Eocene of North America.

dication of any sort of mammalian superiority. Indeed, the extraordinary diversity of birds, along with the fact that a quarter of all mammal species are bats, could just as easily go to show that nonflying mammals have really been missing out for the last 65 million years. Moreover, there are enough giant ground birds on Earth at the present time and in the fossil record to dispel any lingering doubts about the ability of birds to match up to their megamammalian cousins.

The oldest Cenozoic avian behemoths were the diatrymas (Fig. 8.4), huge ground-dwelling birds known from fossil sites across Europe and North America. At 2 meters (6.6 feet) tall with thickly muscled legs, giant claws, and a head the size of a horse's, *Diatryma gigantea* was not a bird to be trifled with. *Diatryma*'s diet has been the source of some controversy over the last few years. The massive head and axe-like beak immediately suggest a predatory lifestyle, but *Diatryma* lacked the hooked upper bill so characteristic of most flesh-eating birds. It also had very stout limbs, which suggests a rather sedate pace of life. These features have led some to suggest that it was a vegetarian, using its huge beak to crop vegetation rather than kill early Cenozoic mammals. However, recent mathematical analyses of the bite-force of *Diatryma gigantea* suggest that if it was indeed a herbivore, its beak was massively overconstructed for the purpose. Compared with all other giant plant-eating birds past and present, *Diatryma* had a beak that was truly titanic, easily strong enough to crush bones. Indeed, the researchers who did the calculations suggested that *Diatryma* may have been a bone-crushing hunter-scavenger much like a modern hyena.

The jury is still out on *Diatryma,* but there is little doubt about the predatory nature of the phorusrhacids, or "terror birds" (Fig. 8.5) that roamed South America from the late Paleocene to the end of the Pliocene (Fig. 3.1) and spread into North America in the late Pliocene and Pleistocene. The dozen or so species of phorusrhacid were 1.5 to 3 meters (5 to 10 feet) tall, lightly built, and obviously faster and more agile than *Diatryma*. The upper beak had the telltale downward-pointing hook of birds of prey, confirming that phorusrhacids were dedicated meat-eaters. They probably ran down fast-moving mammals, seized them with their clawed feet, and used their powerful beaks to tear away mouthfuls of flesh.

It is frequently claimed that these enormous predatory birds rose to prominence in the early years of the Cenozoic because niches for large bipedal predators had been left invitingly open after the extinction of the dinosaurs and because large mammalian predators were not to appear on the ecological stage for another geologic epoch. Leaving aside the issue of whether there are such things as niches for two-legged as opposed to four-legged preda-

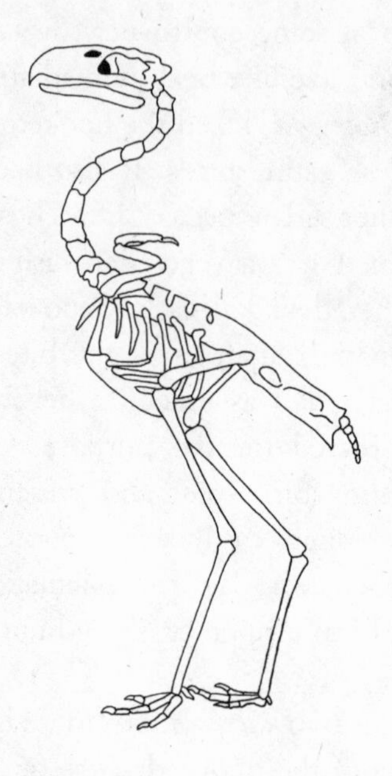

FIGURE 8.5 Skeleton of *Phororhacos*, a type of phorusrhacid, from the Miocene of South America. Height around 1.6 meters (5.3 feet).

tors, there is little evidence to support this rather negative view. Herbivorous and carnivorous mammals were in existence when *Diatryma* walked the Earth; indeed, mammals probably formed the major part of the birds' diet. Although most of these mammals were small, two or three exceeded even *Diatryma* in weight.

The success of phorusrhacids is also often explained away by the fact that marsupials rather than "advanced" placental mammals dominated South America until the land bridge between South and North America became established in the late Pliocene. Once again, there is little evidence to support this view: the notion that placental mammals are more advanced than marsupials—whatever "advanced" might mean—is based on little more than the usual familial prejudice on our part. Many of the doglike marsupial borhyaenids of the time were large and predatory, and no convincing reason has ever been put forward to justify the belief that they were in any way inferior to their placental counterparts elsewhere. In turn, the extinction of phorusrhacids in South America has often been attributed to the southward migration of placental mammals across the newly formed Isthmus of Panama at the end of the Pliocene, but in fact most phorusrhacids passed into the fossil record well before placental carnivores arrived. And the largest phorusrhacid of all, *Titanis walleri,* actually migrated northward at the same time and seems to have made a good living hunting down placental mammals in Florida.

The most straightforward interpretation of the paleontological data currently available is that birds and mammals encountered vacant niches for large predators after the Cretaceous extinction and that birds not only seized the initiative in some parts of the world but held on to it through much of the Cenozoic.

Birds have also produced a number of giant species adapted primarily to a diet of plant material. The most familiar to us today are ostriches, rheas, emus, and cassowaries, all of which subsist on seeds, fruits, green plant material, insects, and perhaps the odd small vertebrate. Cassowaries and emus past and present are known only from Australia and nearby islands, but both groups were much more diverse in the recent geologic past (three species of emus have gone extinct from Tasmania and a number of smaller islands off Australia's southern coast just in modern times). Fossils of rheas are known from as far back as the middle Paleocene of South America and the late Paleocene of Europe, and immense flocks of greater rheas are known to have once roamed the pampas grasslands of South America. Ostriches are now restricted to southern Africa, where they share their environment with a plethora of large herbivorous and predatory mammals, but they are also known from Miocene deposits in Moldavia, and at least four species once ranged widely across the Pliocene and Pleistocene grass-

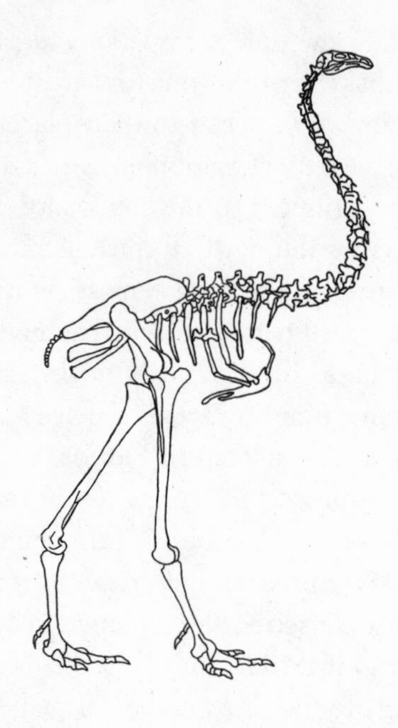

FIGURE 8.6 Skeleton of the recently extinct herbivorous moa *Dinornis maximus* from New Zealand. Height around 3 meters (10 feet).

lands of Africa, China, Mongolia, Ukraine, Kazakhstan, India, and Greece.

Clearly, giant herbivorous birds are able to compete with large mammals. However, the most impressive proliferations of giant avian plant-eaters do seem to have occurred in parts of the world free from significant mammalian competition. The best-known herbivorous megabirds are probably the recently extinct moas of New Zealand (Fig. 8.6), ranging in size from 3.3-meter (11-foot) titans—possibly the tallest birds of the Cenozoic—to species little bigger than a turkey. Their principal predator, at least before the arrival of people, appears to have been another giant bird called *Harpagornis,* the largest known member of the eagle family, with a wingspan of up to 3 meters (10 feet). This exquisitely powerful hunter was probably able to swoop down and break the necks of fully grown moas weighing in excess of 200 kg (440 pounds).

Even heavier than the largest moa was *Aepyornis maximus,* one of seven species of elephant-bird from the Pleistocene of Madagascar. The largest were 2.5 to 3 meters (8 to 10 feet) tall and probably weighed in the region of 400 kg (880 pounds). Australia and New Guinea once had giant herbivorous birds too: mihirungs—an Aboriginal word for "giant emu"—survived in some areas until 26,000 years ago. Typical mihirungs stood about 2 meters (6.6 feet) tall, but one or more species known only from isolated bones may even have rivaled the largest Madagascan elephant-birds in weight. These enormous animals were probably hunted by marsupial wolves and the giant lizard *Megalania,* whom we met in Chapter 6. The idea of a 1-tonne lizard pulling down a ½-tonne bird amid a hail of feathers is the sort of jolt to our familiar notions about the workings of the natural world that only paleontology can provide. The rocks and bones tell us that the world with which we are familiar through our everyday experience is actually quite unusual against the broad backdrop of the Cenozoic.

So, given the available evidence, how should we interpret the relative scarcity of megabirds in Cenozoic ecosystems? If any fundamental physiological or anatomic limitations are involved, they are far from obvious. It is easy to invent hypotheses about the merits of teeth relative to beaks or four legs relative to two, but if diatrymas and phorusrhacids had come to dominate the Earth, then diametrically opposed but similarly convincing post hoc theories would no doubt be spun in their favor.

The most straightforward interpretation is that all Cenozoic birds descended from ancestors that could fly, and evolving a big body necessarily entails abandonment of this defining avian characteristic, whereas all Cenozoic mammals evolved from nonflying ancestors, so growing big required no comparable sacrifice. It is as unsurprising that birds have tended to keep operable wings as it is that mammals have tended to keep their teeth. Some mammals have lost their teeth—anteaters, for instance—but teeth, like wings, are such useful structures that we should not expect animals to give them up very often. As it happens, flightlessness in birds is much more common than toothlessness in mammals, but the overwhelming majority of birds that have lost the power of flight have done so on oceanic islands where the capacity for long-distance travel is not particularly useful and where ground-living predators are usually rare or absent. Time and again under such circumstances the energetic costs of manufacturing and maintaining enormous flight muscles have outweighed the benefits and birds have given up flying. But over most of the Earth's surface, predation by ground-dwelling animals and seasonal cycles of shortage and plenty are pervasive problems against which aerial locomotion is a highly effective foil. The power of flight is a precious gift, so most birds are, and always have been, small enough to reap the benefits.

The scarcity of really big birds is therefore understandable, but there is another puzzle concerning these animals. The proliferation of birds over the last 65 million years has been far in excess of the much vaunted Cenozoic radiation of mammals. Why the difference? Why have birds evolved into new species so much faster than mammals?

The most common weight-class for birds is around 33 grams (1.2 ounces), fully three times smaller than the equivalent for mammals, and it has long been known that taxonomic groups composed of small animals tend to be richer in species than those composed of large ones, probably because the world invariably becomes more complex as animals get smaller. A tree, for example, is little more than a food source to a deer, but it may be home to hundreds of species of insects living and feeding at different levels in the canopy, on the topside or underside of leaves, in fissures in the bark, between the bark and sapwood, amid leaf litter on the forest floor, and so on. Diversity tends to correlate strongly with the number of different ways in which the environment can be exploited, and this number is usually much higher for small animals than big ones. Small creatures also tend to get through more breeding cycles in a given period than large ones. So, all other things being equal, we might expect taxonomic groups composed of small animals, such as birds, to proliferate at a greater rate.

Birds also often inhabit woodlands: complex, three-dimensional environments that offer a correspondingly broad spectrum of potential niches. The highest production of leaves, fruits, seeds, and insects in a woodland is usually high up in the air. Small tree-living mammals do manage to navigate their way through the maze of branches in forest canopies, though their routes are necessarily tortuous and accompanied by the ever-present danger of falling. But birds can flit anywhere unencumbered by gravity and perch on or hang from the thinnest branches and twigs, so they are ideally suited to these high-altitude tangles of immense ecological opportunity.

And birds have proliferated in forests in spectacular fashion. The greatest diversity of birds anywhere in the world can be found in the Neotropical region (Fig. 8.7) consisting of Central and South America, southern Mexico, and the West Indies, and including the immense tropical rain forests of Amazonia. Over 3000 bird species—nearly a third of the global total—occupy this one biogeographic realm, the overwhelming majority in forested habitats. Mammals, on the other hand, show no such bias toward these forests: northern South America holds sixty-six fully terrestrial mammalian genera but only thirty-eight arboreal ones (excluding bats, which we will come to later). The importance of woodland in fostering avian diversity becomes clear when we compare species counts for the Neotropics with the Ethiopian region. Both realms have extensive areas of land in the tropical zone, but the Ethiopian region supports only half the number of bird species.

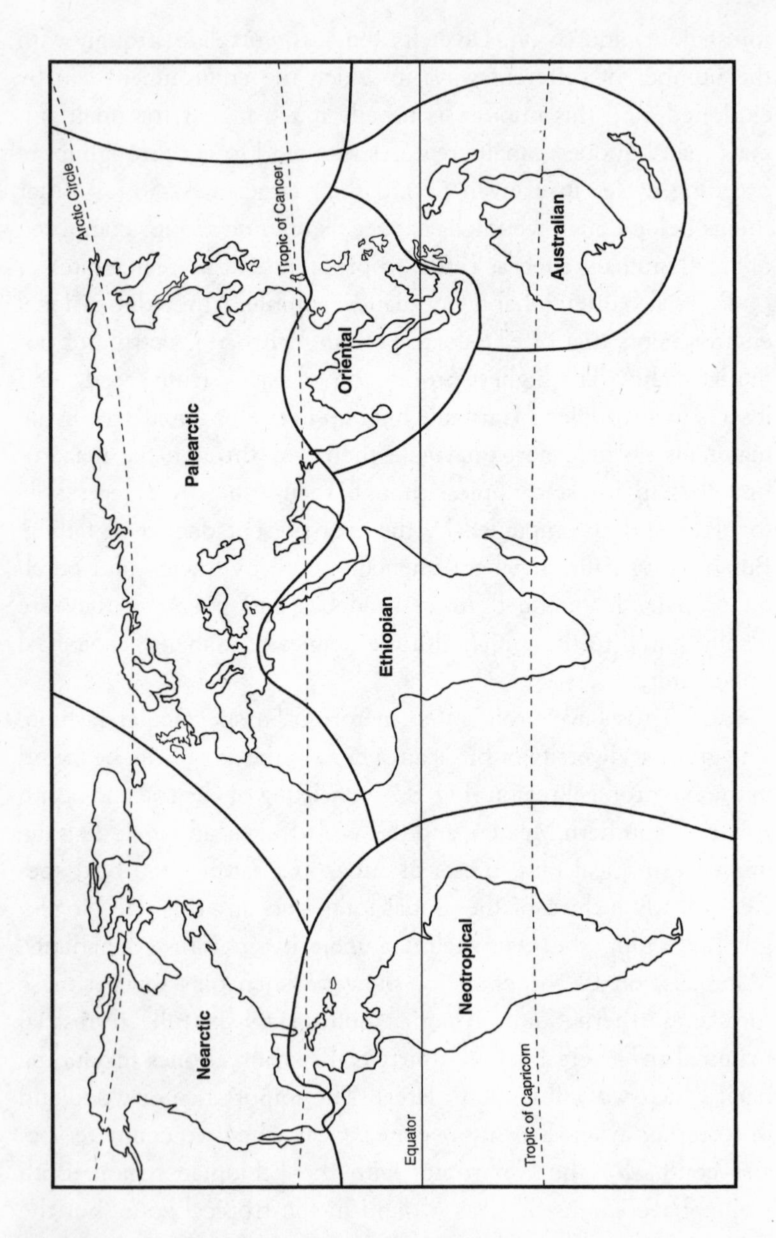

FIGURE 8.7 Biogeographic regions of the world.

One reason for this is thought to be the much greater area of rain forest in the Neotropics. The generally drier climate in much of tropical Africa encourages the development not of forests, but of savanna. Grasslands in general are ideal environments for large, earthbound mammalian grazers and their associated predators, but they are relatively simple habitats in the vertical plane that flying animals are uniquely adapted to exploit.

The correlation of bird diversity with forest area at the continental scale makes sense because geographically extensive habitats are more likely to sample many degrees of latitude, span a wider range of topographic, climatic, geologic, and hydrologic conditions, and thus have a greater variety of subenvironments for the resident flora and fauna to exploit. Strong positive relationships between species diversity and area are extremely common in ecological and biogeographic studies, from the scale of continental landmasses right down to patches of forest or grassland in a landscape. Large patches of woodland, for example, may support viable breeding populations of species with large home ranges (e.g., many birds of prey), whereas smaller areas may not. Small patches of habitat can only support small populations of animals, and small populations are at greater risk of extinction through random changes in population levels or environmental disturbances. And if numerous patches of habitat are spread over a wide geographic area, the risk of all the subpopulations becoming extinct at the same time is much reduced. If a subpopulation does succumb, there is a good chance that the loss can be made good by immigration from somewhere else. All these factors interact—invariably in complex ways that are difficult to disentangle—to influence species–area relationships at broader spatial scales. But although positive species–area correlations are common, the relationship is almost never one to one. That is, a doubling of area usually leads to a much smaller increase in the number of species. This also makes sense because we would not expect a forest to double its species complement as it doubled in area any more than we would expect the number of nationalities among the residents of a ten-story block of apartments to be twice as high as in an adjacent five-story one. We might expect to find a few more nationalities in the larger block just because we have sampled more rooms, but it would be a huge statistical surprise if the number of nationalities turned out to be double.

The number of bird species in a given area of Neotropical forest also tends to be much higher than in Africa, probably because of differences in forest history. Cycles of relatively dry climate associated with the ice ages broke up tropical forests into fragments separated by wide areas of savanna.[5] Because the flow of genes between each fragment was restricted, the likelihood of separated populations evolving in different directions increased. With the return of wetter conditions during glacial interludes, the forests expanded and became connected again, as did the geographic ranges of many of the resident birds. Many populations had diverged from their neighbors during isolation to the point where they could no longer interbreed, and the result was a greater number of species in the forest as a whole. Although this process of periodic isolation and reconnection is thought to have happened in both Neotropica and Africa, the number of forest refuges has been estimated at around twenty-seven in the former but only six in the latter. More refuges means more opportunities for divergence and speciation, so it is no surprise that South American rain forests have ended up with a higher density of bird species.

Bats, the only other group of flying warm-bloods, show very similar patterns of diversity. These animals are very strongly concentrated in tropical regions, with hot spots in heavily forested

---

5. Like most paleoenvironmental reconstructions, Quaternary refuge theory is controversial. See Bush and Colinvaux (1990) and Williams et al. (1998) and references therein for alternative theories and detailed discussions.

parts of the Neotropical and Oriental regions. But at similar latitudes in tropical Africa, species density is two to three times lower. James Findley has done some interesting statistical analyses in an attempt to explain these patterns. He recorded the number of bat species, measured the area of rain forest, and estimated the number of refuges in five regions—Africa, mainland southeast Asia, Neotropica, Borneo, and New Guinea. (The refuges in Neotropica and Africa are postulated glacial ones, whereas the rain forests of the Indo-Australian archipelago are divided into natural fragments because they exist on islands separated by stretches of ocean.) He found a strong correlation between the number of bat species and rain forest area, and when he added in the values for the number of refuges, the correlation became almost perfect. As Findley himself is quick to point out, there is considerable room for error in estimates of the number of refuges in each region, and one should always be aware that correlation does not necessarily imply causation, but the potential influences of forest area and refuge history are so obvious that we can be pretty confident that the statistics are telling us something important about the factors controlling global patterns of bat diversity.

So the great proliferation of flying warm-bloods in the Cenozoic can be put down in part to the history of forested environments, the broad spectrum of ecological opportunities offered by them, and the advantages of flight in a three-dimensional living space where the chief concentration of resources is high up in the air. The unique ability of birds and bats to exploit woodlands is probably the main reason for their diversity, but the power of flight also presented Cenozoic aeronauts with another major arena for proliferation from which ground-living mammals have been largely excluded. We have already seen that connectedness is an important feature of the ocean realm from the point of view of seafaring mammals, but the most connected of all earthly environments is the atmosphere, which stretches unbroken and unobstructed around the entire globe. Unlike earthbound creatures, therefore, animals capable of traveling the airways potentially have access to all the terrestrial environments on the planet.

And there have been a lot of far-flung places on offer in the Cenozoic. About 125 million years ago a plume of molten rock detached itself from the Earth's outer core and rose up like a gigantic balloon beneath what is now the western portion of the Pacific Ocean. As it rose, the pressure from the overlying rocks gradually lessened and the more volatile components boiled to the surface, erupting through the sea floor as volcanoes, which eventually spread over much of the Pacific basin. From the mid-Cretaceous onward, some of these immense accumulations of rock grew so high that their summits broke above the waves to form the major island groups of Polynesia, Melanesia, Micronesia, and the Malay archipelago. Similar geologic processes are forming islands in the Pacific today, such as Hawaii, whose history as a piece of land stretches back a mere 6 million years. Exposed to the wind, lashed by storms, and pounded by waves, oceanic islands are quickly worn away, but for a short while they become available to any terrestrial plant or animal capable of crossing the sea and setting up house.

Animals and plants get to islands stuck out in the ocean in a number of ways. Birds and bats actively fly and in the process transport the seeds of fruit-bearing plants in their guts and the eggs of snails and other creatures attached to their plumage. Insects, spiders, and seeds may arrive on the wind. Coconuts and the seeds of some other coastal plants float in. If the island is close to a continental source area and in relatively shallow water, larger animals may be able to walk over during periods of low sea level. Smaller animals may get washed down rivers after storms and float to an island either suspended in the water or perched on a raft of floating debris. Which of these dispersal mechanisms is most important for nonflying tetrapods has been the cause of much dispute among scientists in the past. There have been long-running arguments, for example, about how lizards colonized islands in the Caribbean: did they spread to their current locations over water or were there land bridges or closely spaced stepping-stones connecting the islands with continental America in the past? The proponents of land bridges usually contend that long-distance dispersal over water is such an unlikely scenario as to be not worth considering, while others, usually geologists and paleontologists, who are used to imagining vast expanses of time, point out that over many millions of years extremely unlikely events are, in fact, quite likely to happen. The dispute has stayed alive because neither camp has had the opportunity first-hand to study patterns of dispersal during periods of low sea level, nor animals colonizing islands on rafts of debris.

Not, that is, until recently. In 1998, Ellen Censky and her colleagues reported an irrefutable example of over-water dispersal by a breeding population of large reptiles. The animals were green iguanas, hefty, thickset lizards up to 40 cm (16 inches) in length (excluding tails). At least fifteen individuals washed up on a beach on the island of Anguilla on 4 October 1995 along with the remnants of their raft of logs and trees. Anguillan fisherfolk reported that the raft was so big that it took two days to pile up on shore. They also saw the iguanas both on the beach and perched on logs in the bay. Exactly one month earlier, hurricane Luis had ripped through the Caribbean, followed a couple of weeks later by the smaller hurricane Marilyn. From the tracks of the hurricanes and the distribution of green iguanas in the Lesser Antilles before 1995, Censky suggests that the animals floated all the way from one of the islands of Guadeloupe some 300 km (186 miles) to the southeast. Depending on which hurricane rudely dislodged the iguanas from their previous home, they must have clung precariously to their logs for between two weeks and a month. In March 1998 the iguanas were still on Anguilla and one of the females appeared to be pregnant. So over-water dispersal and colonization of islands by breeding populations of large animals is clearly possible. It must be an unlikely event, certainly, but events that appear improbable to our abbreviated consciousnesses will happen quite often over the course of millions of years.

The colonization of distant oceanic islands by reptiles must be largely if not entirely the result of passive dispersal either in the water or on rafts of some kind. The same must be true for non-flying land mammals. But there is one fundamental limitation on the dispersal capabilities of terrestrial mammals: their thirsty metabolic engines. Rats and cats are just as likely to get washed out to sea or marooned on floating debris as lizards and snakes, but whereas a reptile could conceivably survive passively in the water or hide itself somewhere on a raft for weeks or months, mammals without access to food or fresh water would be dead within hours or days (longer for larger animals, but the larger the animal, the less likely that it would be transported in the first place, let alone a viable breeding population). An animal may be able to find some food on a raft, but any that is available would certainly sustain a reptile much longer than a mammal.

These simple differences in transportability and endurance result in the sort of over-water colonization abilities illustrated in Figure 8.8. Tiny animals disperse easily on the wind, bats and birds power themselves across the oceans, and snails hitchhike as eggs attached to flying animals of one sort or another. All these methods allow long distances to be covered across wide tracts of inhospitable ocean. Lizards raft considerable distances across the sea aided by their fuel-efficient engines. Tortoises drift to distant islands in much the same way that coconuts do. Nonflying mammals, however, especially large ones, are very poor transoceanic travelers. Rodents appear to be the best of a poor bunch: rats and mice are small, they often exist in enormous numbers, and they tend to be rather catholic in their eating habits. This combination of factors means that they may occasionally raft across stretches of ocean and find some way of eking out a living in a new island

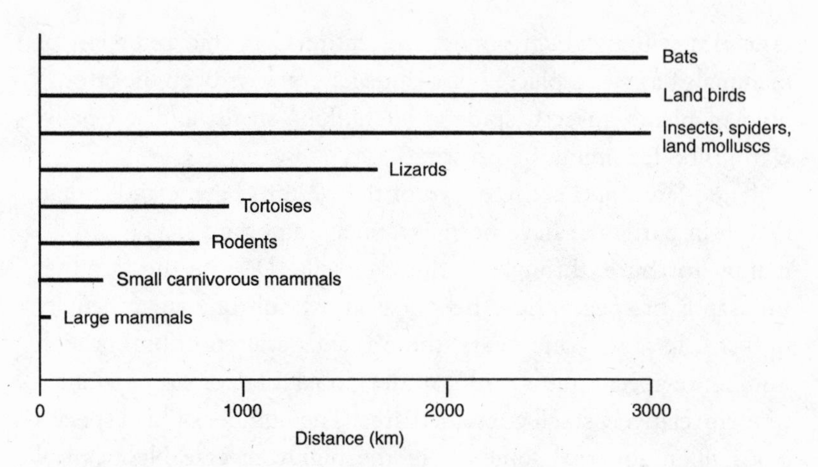

FIGURE 8.8 Maximum known over-water dispersal distances for various groups of animals (excluding those transported by humans). Adapted from Gorman (1979). (1000 km = 620 miles).

environment. Rats in particular are probably the best nonflying mammalian travelers, with many endemic species (those found in a particular place and nowhere else) on islands in the Pacific and many other archipelagos around the world. But the general pattern of colonization abilities shown in Figure 8.8 is clearly reflected in the biogeography of the major groups of animals and plants on Earth: ancient[6] continental landmasses have been the evolutionary

6. Continents are thick accumulations of relatively light rock that form tectonic plates in conjunction with areas of denser oceanic crust. Plates move because they are pushed from linear spreading zones in the middle of ocean basins where molten rock rises to the surface to form new strips of sea floor. When plates meet, a number of things can happen depending on the type of crust on either side of the suture. At ocean-ocean boundaries, one plate is pushed down under the other and melts. The molten rock rises to the surface to form volcanoes which, if effusive enough, eventually become oceanic islands. At ocean-continent boundaries, oceanic crust is pushed under because it is more dense. At continent-continent boundaries, two thick slabs of buoyant material come together and neither can be pushed down. This is what happened when India hit the continental portions of the Chinese and Eurasian plates, forming the Himalayas. Thus, oceanic crust is constantly being eaten up at plate boundaries and replaced at

crucibles within which almost the entire Cenozoic radiation of mammals has taken place, but islands have always been the stronghold of plants, insects, spiders, hitchhiking snails, and, crucially, warm-blooded animals with wings.

The islands and archipelagos of the western Pacific and Indian Ocean in particular have been extremely important sites of proliferation for birds throughout the Cenozoic. Despite the fact that thousands of species have been lost as a result of human activity in just a few thousand years, these islands still contribute a high number of species (about 730) to the global total, disproportionate to their relatively small combined area. The number of bird species per million square kilometers in the highly diverse Neotropical region, for example, is around 177, while the Malagasy region, dominated by the large continental island of Madagascar but also including the western Indian Ocean islands, contains 328 species per million square kilometers. But with a staggering 4894 indigenous species per million square kilometers, the volcanic islands of the tropical Pacific are clearly out of the ordinary. Care must be taken with this sort of comparison, because the density of species is liable to be higher in smaller regions simply because diversity tends to rise slowly with increasing area, but the extent to which endemic birds are packed onto islands in the tropical Pacific nevertheless stands out as highly unusual.

Unusual, but not mysterious. The Pacific islands lie mostly at tropical latitudes and, as we saw in Chapter 5, diversity tends to increase toward the equator, but the main reason for the high

---

spreading centers, which means that it tends to be relatively young. Any seamounts or islands pockmarking the surface will necessarily be young too. Buoyant continental crust, however, has been on top throughout geologic history and is thus extremely old. Over billions of years, continental landmasses have been inundated by the sea, covered with sediment, eroded by wind, waves, rivers, and ice, jammed together and split apart, but they have always formed the permanent stage on which the long-running drama of terrestrial evolution has been played out.

number of indigenous birds in the region is the division of the constituent land areas into small chunks separated by tracts of ocean. Just as in climatically isolated rain forest fragments, natural selection and genetic drift gradually modify island populations according to their genetic inheritance and environmental circumstance. Given sufficient time, the populations may diverge until they would no longer be able to interbreed even were they to meet, at which point one species has become two.

Compounding the isolation effect, oceanic islands usually offer a wide range of ecological opportunities because only a small subset of the continental species pool manages to reach them. A colonist of a new island forest is in a completely different situation from its relatives back home, because continental forests are already full of species that have been exploiting and adapting to them for millennia. There may be woodpeckers, nuthatches, and treecreepers picking food off tree trunks, tits and goldcrests taking insects high in the branches, flycatchers hunting on the wing, wrens rooting about in the understory, and so on. A finch in a continental forest would have little chance of gradually adapting to a woodpecker's way of life for the simple reason that continental forests are already full of actual woodpeckers who are much better at it. And all the other niches are equally firmly occupied. But a wayward finch landing in a remote island forest is likely to find many, if not all, of these birds absent. There may be enormous numbers of insects and grubs going begging on tree trunks, and any attempt to harvest them, however inept initially, is likely to reap rewards. Given enough time free from significant competition, the ancestral species may give rise to bark-gleaning, fruit-eating, fly-chasing, nut-cracking, and nectar-sipping species that would never get a chance to evolve elsewhere. Where once there was a single species of continental affinity, suddenly (in geologic terms) there are numerous species all adapted to different ways of life and all unique to the archipelago in question. This sort of proliferation is called "adaptive radiation," and examples from distant islands are often described as "explosive."

The honeycreepers of Hawaii and its adjacent islands are a good example of such an evolutionary explosion. The Hawaiian chain of volcanic islands lies 3200 km (1984 miles) from California to the east and a similar distance from the Marquesas to the south, so very few animals have ever managed to get there. Hawaii has no native reptiles, amphibians, or freshwater fish and only one native land mammal, which, not surprisingly, is a bat. However, at least five species of thrush evolved on the islands, each occupying its own island or island group. Small flycatchers called elepaio, found only in the Hawaiian islands, have given rise to distinctive subspecies on Kauai and Oahu and a further three on Hawaii itself. But the most remarkable birds of the islands are without doubt the honeycreepers. The ancestor of the group was probably a finchlike bird from Asia. How a viable breeding population managed to reach the islands across such a vast expanse of ocean we will never know, but it would not be at all surprising if a hurricane was involved. From this original stock at least fifty species of what we now call honeycreepers evolved, half of which still occupy the islands. The remainder are known only from fossils, most having become extinct since the arrival of people in around A.D. 300.

The initial colonists probably fed on nectar and insects, and from this starting point the great variety of honeycreepers with their varied diets and beaks evolved (Fig. 8.9). Many of the extant genera, such as *Himatione, Vestiaria, Palmeria, Drepanidis,* along with many species of *Loxops* and *Hemignathus procerus,* still feed on nectar. Insects are often attracted to this rich food source too, and most nectar-eating honeycreepers supplement their diet with any insects that happen to present themselves. A number of species have since made the transition to an insect-only diet: *Hemignathus wilsoni* uses its long upper beak as a probe and its shorter lower bill to lever insects out of crevices in the bark of trees, while *Pseu-*

FIGURE 8.9 The remarkable adaptive radiation of the Hawaiian honeycreepers. 1. *Ciridops anna*; 2. *Palmeria dolei*; 3. *Himatione sanguinea*; 4. *Vestiaria coccineo*; 5. *Drepanidis pacifica*; 6. *Loxops coccinea*; 7. *Loxops virens*; 8. *Loxops parva*; 9. *Hemignathus procerus*; 10. *Hemignathus lucidus*; 11. *Hemignathus wilsoni*; 12. *Pseudonestor xanthophrys*; 13. *Psittacirostra bailleui*; 14. *Psittacirostra cantans*; 15. *Psittacirostra kona*; 16. *Psittacirostra psittacea*. Adapted from Cox and Moore (1993).

donestor xanthophrys from the forests of Maui uses its much stouter bill to rip open twigs and branches in search of wood-boring beetles. Other *Hemignathus* species use their beaks as hammers, like woodpeckers, to drive insects out from under bark. Species of the genus *Psittacirostra* evolved heavy, powerful beaks which they use for cracking open seeds and nuts. The recently extinct *Ciridops anna* used its much lighter bill to feed on the soft fruits of Hawaiian palm trees. All this riotous diversity derived from a single species blown over 3000 km across an inhospitable ocean and helped along by the ecological opportunities presented by the simple absence of others.

This sort of explosive diversification has been happening on islands all over the world throughout the Cenozoic. More than 1750 bird species—18 percent of the current global total—are confined to these relatively tiny areas of land sticking out of the sea. This is more than 40 percent of the total number of mammal species on the entire planet. And the proportion would have been much higher in the past. The impact of humans on island ecosystems—and thus on birds in particular—has been profound: 85 percent of bird extinctions in historical times have occurred on islands, a rate forty times higher than on continental landmasses. The extent of human-caused mammal extinctions on islands is unknown, but given that nonflying mammals hardly ever reach islands anyway, the figure must be minuscule in comparison. We will return to the subject of islands and the effects of human colonization in the next and final chapter.

So the marked disparity between the number of mammal and bird species on Earth has a number of probable causes. For a start, birds are on average much smaller than mammals, and diversity tends to rise as body size decreases across a wide range of taxonomic groups. The environment provides a greater variety of places for small animals to build nests and raise their young, forage for food, and escape from competitors, predators, and the vicissitudes of the physical environment. Diversity of opportunity and diversity of species are tightly linked, and the opportunities for small animals in a complex world are simply much greater than for large ones.

But the power of flight has probably been the key factor. The way in which humans perceive the environment is necessarily constrained by our huge bulk and our one-way relationship with gravity. We walk through forests and see mostly tree trunks. We can't even look at the canopy without risking a collision or at least a cricked neck. And our world is full of barriers: mountain climbing is such a feat of endurance that its practitioners often gain world renown, and swimming the English Channel is still an endeavor worthy of news coverage. But to many birds, stretches of ocean and mountain ranges are little more than different types of scenery. The world of opportunity from which we and our sedentary mammalian cousins have been excluded over the last 65 million years is vast.

From the tallest trees in the most remote regions of Amazonia to the tiniest, loneliest scraps of land in the middle of the Pacific Ocean, one can always be sure of finding feathered dinosaurs. But the avian empire is not just limited to treetops, remote islands, and other places free of mammalian competition. Birds exploit almost every type of terrestrial and marine environment, from the refrigerator of the Antarctic to the deserts of the Sahara. And they have even taken on mammals at their own megafaunal game, with remarkable success given the fundamental restructuring of the avian body plan to accommodate an aerial existence. With no teeth and only two sets of claws, the diatrymas and phorusrhacids still managed to dominate as top predators in the Americas for much of the Cenozoic. Mihirungs flourished in the worrisome company of marsupial wolves and the largest predatory lizards in history. Ostriches share their present-day African environment not only with large plant-eating mammals but with some of the scariest mammalian predators anywhere on Earth. Certainly terrestrial megabirds are rarer than their mammalian counterparts, but this is not surprising. There are, after all, no ground-dwelling megabats at all. Nor is there any evidence of earthbound pterosaurs in the fossil record. The likely explanation in all three cases is simple: most animals with wings have better things to do.

[9]

TWO WARNINGS FROM HISTORY

There is a natural order to tetrapod life on our planet, rooted in the natural ecological talents of different types of animals that we have been exploring in this book. Because of their fuel-efficient metabolic engines, reptiles have enjoyed unchallenged dominion as large animals in freshwater for most of tetrapod history. The poor and unpredictable arteries of the land remain as alien and dangerous to big warm-bloods today as they must have been to the mighty dinosaurs of the Mesozoic. Amphibians do not travel well across the sea because neither they nor their eggs fare well in saltwater, but as small animals in cool, wet places on continental landmasses, these delicate-skinned creatures hold their own very successfully, just as they have always done. In drier places, armies of tiny, flat, or crack-shaped reptiles excel at hiding and waiting in the nooks and crannies of terrestrial environments where big, greedy warm-bloods find it difficult to follow, while others find ecological opportunities far from the frenetic mammalian hordes by rafting across vast tracts of ocean with their lean-burn engines barely ticking over. Cold-bloods may be inveterate hiders and plodders, but as experts in the ecological strategies of frugality and physiological durability, nothing else in the tetrapod sisterhood can hold a candle to them.

In contrast, with our elegant temperature-control systems and adaptable insulation, warm-bloods go unchallenged as large animals at the thermal ends of the Earth. Only we can stand exposed to the Saharan sun and the Antarctic chill and tough it out. And our high-capacity aspirated engines give us the power, sustained speed, and physical endurance to outpace, outdistance, and outcompete our largest cold-blooded rivals on land in all but the poorest, most unpredictable places. Without sustenance our fundamental physiological weaknesses become apparent very quickly, but in places of plenty with nowhere to hide, we warm-bloods are rugged enough to bully our way to the top almost every time. Our insatiable need for food and fresh water also imprisons most of us on the continental landmasses of our evolutionary birth, but with ultralight bodies, wings, and phenomenally powerful metabolic engines, two of our number—birds and bats—have managed to claim outposts on even the most remote oceanic islands. For 65 million years we survivors of the Cretaceous asteroid impact have played to our talents, worked around our shortcomings, and divided the world up between us.

But the natural order of tetrapod life is not governed by immutable laws. Large flightless birds are atypical, but they do exist. Flightless bats are unknown, but no one would be very surprised if one turned up. Mammals are not good seafarers, but rodents do manage to reach distant islands from time to time. Big land animals are nearly always warm-blooded, but *Megalania* reaches out from the fossil record to remind us that even the strictest rules can be broken. Desert amphibians are clearly ludicrous, but there are spadefoot toads. All these charming exceptions to the rules are either realized or easily imagined. They neither contravene the laws of physics nor lie so far along the bell curve of probability as to be completely incredible. But what are the chances of one day finding a flying animal five times heavier than the heaviest known? Or of a breeding population of 50-kg-plus warm-bloods traveling 3000 km across the Pacific Ocean on a raft of floating vegetation?

Under the normal rules of tetrapod life the probability in both cases is not significantly different from zero. So next time you look in the mirror, consider how remarkable you really are.

Of course, it took some working out, particularly the flying part. But even long-distance rafting is such a flagrant violation of the natural order for big warm-bloods that it required careful thought and planning. Vegetation had to be lashed or nailed together to keep it in one piece, then stocked with enough food and water to keep a breeding population of large creatures with outlandish appetites alive for the duration of any transoceanic endeavor. Drifting around at the mercy of ocean currents in the hope of making landfall is an excellent way of running out of fresh water, so a propulsion system—oars or sails—and a steering mechanism were added. With intelligence and ingenuity, humans have been rafting around the oceans for over 50,000 years, but in particularly high numbers for just the last 4000,[1] all in direct contravention of a fundamental rule of life that has been firmly in place for at least 200 million years.

And it is largely because big mammals are not traditional seafarers that the impact of itinerant humans on other animals around the world has been so profound. Until very recently the avifauna of New Zealand included moas, huge eagles, pelicans, swans, giant ravens and flightless ducks, coots, and geese, none prepared by their evolutionary history to cope with the presence of hairy predators. Within a few centuries of our arrival all were extinct. A coincidence? "[A]n incredible [one] if every individual of dozens of species that had occupied New Zealand for millions of years

---

1. Australia and New Guinea were colonized ca. 50,000 and 40,000 years ago respectively. Humans reached New Ireland and the Solomons 29,000 years ago but then paused for 25,000 years, probably because more distant islands were undetectable. Then the ancestors of Polynesians rafted right across the Pacific between 2000 BC and AD 1000. Nearly all the inhabitable islands of Polynesia, Melanesia, and Micronesia were colonized by humans at some point during this latter period.

chose the precise geological moment of human arrival to drop dead in synchrony."[2]

Archaeological evidence suggests that the first colonists of New Zealand rapidly exterminated moas by hunting, nest-robbing, and forest clearance. The other species probably succumbed for some or all of the same reasons. Similarly, before the arrival of people, Madagascar sported elephant-birds, two species of giant tortoise, a dozen lemurs up to the size of a gorilla, an aardvark, a pygmy hippo, and a giant mongoose. All are now extinct.

A full list of these island extinctions would make tedious and depressing reading. Suffice it to say that had we not rafted across the Pacific, the world would be about 20 percent richer in bird species alone.

Much of this devastation, particularly of larger species, was brought about by hunting and habitat destruction, but just as important—probably more so in the long run—has been the liberation of other organisms from the bonds of the natural order. There was a time when hitchhiking across oceans attached to other creatures was the particular talent of seeds and snails. These organisms are still using their traditional vectors as well as our transoceanic vehicles, but in the last few thousand years rats have started doing it too. And cats. And even cows. Never in Earth's history has a breeding population of cow-sized animals rafted across an ocean, but in a geologic microsecond humans have changed the rules. The ancient geographic barriers that have long restricted warm-blooded landlubbers to their continents of origin are breaking down, and we are directly responsible.

Exclude ground-dwelling mammals from a community for any length of time and other creatures tend to let down their guard. Island plants, for example, often lose their defenses against grazers: the benefits of manufacturing thorns, poisons, and tooth-eroding

---

2. Diamond (1992).

granules of silica may outweigh the costs where grazing animals are commonplace, but on mammal-free islands there may be no benefits at all. Similarly, on islands free of predatory rats, cats, foxes, mongooses, and weasels, birds may have an energetic incentive to give up flying: flight muscles, after all, are extremely expensive organs to manufacture and maintain. And so, over time, the genetic memory of aggressive ground-dwellers begins to fade and island residents drift into distinctly uncontinental ways of life.

Because mammals are often the dominant ground-dwelling creatures on continents and usually the most obvious absentees from islands, displaced mammals have caused more problems for island communities in the last few thousand years than any other type of tetrapod. Particularly ruthless in exploiting the ecological opportunities associated with transplantation is a creature that continental humans tend to regard as rather inoffensive: "We cannot discuss the ecology of islands without making a few disparaging comments on goats. These creatures must be the true embodiment of the devil for a plant lover."[3] It is rare for scientists to talk in such emotive terms, but goats provide ample justification. The natural range of the wild goat extends from Asia Minor through the Caucasus and southern Turkmenia to Iraq, Iran, Beluchistan, and India, but domestic goats have been carried around the globe by us and are now the most widely naturalized feral mammals in the world. They can be found in Britain, North and South America, Australia, New Zealand, some Mediterranean islands, Bonin island in Japan, the Galápagos and Hawaiian islands, Fiji, Saint Helena, Mauritius, the Seychelles, and numerous smaller patches of oceanic real estate right around the globe.

Goats can thrive in areas too poor to sustain viable breeding populations of other herbivores, and they are remarkably tolerant

---

3. Koopowitz and Kaye (1990).

of the oily and bitter-tasting substances manufactured by many plants to discourage the attentions of grazers. They browse from shrubs and trees and have even been known to climb them to reach the most succulent foliage. They gobble up any mast or fruit lying on the ground and quickly behead any seedlings that have the temerity to emerge above the surface. Because these mobile waste-disposal units can survive on almost any type of organic material, they simply continue to eat whatever is available until their adopted landscape begins to resemble a moonscape.

The great Victorian traveler and naturalist Alfred Russel Wallace described in 1895 the devastating effects of goats on the island of Saint Helena in the South Atlantic:

> When first discovered, in the year 1501, St. Helena was densely covered with a luxuriant forest vegetation, the trees overhanging the seaward precipices and covering every part of the surface with an evergreen mantle. This indigenous vegetation has been almost wholly destroyed and . . . the general aspect of the island is now barren and forbidding. . . . [W]hen [the vegetation] was destroyed, the heavy tropical rains soon washed away the soil and left a vast expanse of bare rock and sterile clay. This irreparable destruction was caused in the first place by goats, which were introduced by the Portuguese in 1513 and increased so rapidly that in 1588 they existed in thousands. . . .
>
> In 1709 the Governor of the island complained that the timber was rapidly disappearing. . . . In 1809 the Governor reported the total destruction of the forests. . . .

Today, the native vegetation of Saint Helena survives only on elevated ridges and in narrow gullies. Of thirty-three plant species endemic to the island, goats have eliminated ten and endangered a further eighteen.

And it has been a similar story in other parts of the world.

Grazing, browsing, and trampling by goats devastated the vegetation on Santa Catalina, causing sharp declines in the endemic subspecies of the California quail, the Channel Island gray fox, Slevin's deer mouse, the western harvest mouse, and three species of snake. Similar degradation by goats led to the extermination of the Santa Barbara song sparrow and the Pintan race of the Galápagos giant tortoise. Seventy-three of 143 plant species on Great Island off New Zealand were exterminated by goats just between 1890 and 1934. And because of depredation by alien goats and rabbits, Round Island in Mauritius now has the distinction of supporting more threatened species of plant and animal per hectare than anywhere else on Earth.

Goats also threaten the flora and fauna of Australia, but they are not as much of a problem as rabbits. Some 2300 years ago these creatures were restricted to the Iberian peninsula and perhaps northwestern Africa, but they have since traveled with humans around the Mediterranean, through most of Europe, down to southern Africa, across to America, Chile, Australia, New Zealand, and are now resident on 550 islands from Hawaii to the Falklands. Most notoriously of all, twenty-four rabbits were shipped to Australia from England by Thomas Austin in 1759 and released on his estate as game. Only 120 years later the rabbit population of Australia stood at 700 million. "Of all the introduced pests," wrote one commentator, "the rabbit is by far the worst. Its acclimatization in Australia was the greatest single tragedy that the economy and the native animals ever suffered."[4] Intense grazing, particularly in the semiarid parts of Australia, has decreased the amount of food available for both domestic stock and the indigenous grazing marsupials. Rabbits also inhibit the regeneration of grasses, which in turn decreases the supply of food for granivorous birds. Overbrowsing of shrubs and ring-barking

4. Frith (1979).

of trees has reduced the cover of vegetation required by many indigenous birds for shelter and nest-building. In many areas the combination of these activities has decimated vegetation cover and caused serious soil erosion.

Attempts to control rabbits in Australia have met with limited success and provide a salient warning of the dangers and costs of combating introduced species. Between 1883 and 1892 a series of fences were constructed, some over 1000 km (620 miles) long, in an attempt to stop the march of rabbits across the Australian landscape. They were effective for a while, but the combined effects of floods, high winds, and unobservant emus and kangaroos soon gave rabbits free passage again. Shooting, netting, gassing, and poisoned carrots were all tried. Unfortunately, native marsupials eat poisoned carrots too: combined with the effects of drought, poisoning led to the actual or near extinction of bridled nailtail wallabies, rabbit-bandicoots, Gaimard's rat kangaroos, brown hare-wallabies and serious declines in the populations of many other species. The introduction of the myxoma virus in 1950 killed an estimated 98 percent of Australia's rabbits, but by 1955 the survivors were already staging a comeback. The virus has remained active since its introduction and has kept rabbit numbers below pre-1950 levels, but by the early 1990s the recovered population was still costing the country over 100 million Australian dollars per year in agricultural losses alone. Consequently, in 1996, rabbit calicivirus was released into the population. Over 80 percent of the rabbits in some arid areas contracted the virus and died, and state-funded programs to keep down the remainder by poisoning and mechanically "ripping" warrens have been implemented in some regions. The effect of the disease on population numbers appears to vary considerably depending on climatic conditions, however, and the evolution of some level of resistance to the virus is inevitable, so the long-term impact of this control measure is uncertain. What *is* certain is that even the combined effects of myxoma, rabbit calicivirus, poisoning, and warren destruction will not eradicate these animals completely. Will rabbit numbers now remain under permanent control and allow affected native species to recover? No one knows, but my money is on the visiting team.

Another attempt at rabbit control involved the breeding and release of introduced feral cats. Unfortunately, they had little effect on the rabbit population. Would that the same could be said for native Australian animals. Cats prey particularly heavily on possums, numbats, planigales, and marsupial mice and are thought to have been responsible for the near extinction of ground parrots, noisy scrub birds, and, with the help of introduced red foxes, the serious decline of brush-tailed rock wallabies. They continue to prey on the one surviving population of the critically endangered eastern barred bandicoot.

But the impact of cats in Australia is nothing compared with the devastation they have wrought in New Zealand. Apportioning blame for declines in native New Zealand wildlife is never straightforward because the impacts of introduced cats, three species of rats, stoats, weasels, ferrets, imported diseases, and habitat loss are understandably difficult to disentangle, but the activities of cats are strongly implicated in many instances. On the main North and South Islands of New Zealand, cats have contributed to the decline of kakapos, bush wrens, bell birds, parakeets, New Zealand shelducks, collared gray fantails, New Zealand stilts, wekas, pheasants, quails, long-billed plovers, kiwis, shearwaters, petrels, penguins, terns, gulls, short-tailed and long-tailed bats, and some lizards. On Mangere Island cats were released to control rabbits but rapidly exterminated the endemic robin and the Chatham Island fernbird. On Chatham Island itself, cats appear to have been the main cause of the extinction of the local island rail and have since begun harrying the endangered magenta petrel. On Stewart Island they have exterminated fernbirds, subantarctic snipe, tuis, robins, yellowheads, saddlebacks, kokakos, and bush wrens. On Herekopare

Island they have wiped out subantarctic snipe, banded rails, New Zealand creepers, fernbirds, robins, and yellow-fronted parakeets. A single cat belonging to the keeper of the Stephen's Island lighthouse famously eradicated a rare endemic lizard and the Stephen's Island bush wren.

And cats, goats, and rabbits are just examples of the global problem of hitchhiking mammals. The cast of displaced villains includes Australian brush-tailed possums, wallabies, and European hedgehogs in New Zealand; horses and donkeys in places as remote as the Galápagos and Hawaiian islands; Arabian camels in Australia; water buffaloes in South America, Indonesia, New Guinea, and Australia; gray squirrels in Britain; Canadian beavers in Finland and the former USSR; muskrats in South America and right across Eurasia; South American coypus in North America, Eurasia, Africa, and Japan; Central and South American Virginia opossums on various Caribbean islands; northern red foxes and feral dogs in Australia; stoats, weasels, and ferrets in New Zealand; North American mink across Eurasia; small Indian mongooses in Indonesia, Tanzania, South America, and various islands in the Caribbean, Hawaii, and Fiji; and cattle, pigs, sheep, deer, Polynesian rats, black rats, and brown rats right around the globe.

In some parts of the world different mammals have invaded in waves, leaving multiple trails of devastation behind them. Torbjörn Ebenhard (1988) relates a typical antipodean example:

> On an unfortunate day of 1919 the captain of SS *Makembo* stranded his distressed ship on the beach of Lord Howe Island, off the east coast of Australia. The fauna of this island had never before encountered a mammalian predator. Now the ship rats streamed ashore, invaded the humid forests, and proliferated totally unchecked. During the following 20 years the rats exterminated five endemic passerine birds. The number of lizards, land snails and other invertebrates decreased due to the rats, and

> sea birds were restricted to breeding on small rat-free islets. Another endemic bird, the Lord Howe wood rail . . . still exists but is seriously threatened by rats through egg predation and its habitat is shrinking due to the activities of goats and pigs. The pigs damage the forest floor through rooting, while goats induce new habitats often invaded by exotic plants, through browsing and grazing. Reproduction among the endemic palms is seriously disturbed by the goats, which consume all shoots, and by rats which predate on seeds.

And these are just some of the effects of a few of the more notorious mammalian emigrants. In our travels over the last few thousand years we have transplanted many amphibians and birds and hosts of invertebrates and plants. Although displaced mammals provide the most dramatic and fast-acting examples of ecological change, it is only in the last few years that scientists have begun to appreciate that the most insidious threat comes not from particular species but from the slow stirring of the biosphere as a whole. Botanists, for example, are now finding that oceanic islands across the globe are beginning to look depressingly similar. Over the last few thousand years the same trees have been planted and the same alien plants introduced. We have taken sacks of our most productive agricultural grains and fruits on our journeys, and agricultural weeds and insect pests have hitched a ride too. Species unique to particular islands become aliens if they are transported elsewhere, so even random stirring of the Earth's biota as a result of our wanderings necessarily increases the prevalence of alien invaders and the problems associated with them.

Island ecosystems are particularly vulnerable to aliens, but continental landmasses are under siege too. For twenty-five years the Nature Conservancy of North America has been compiling a list of some 6500 species on American soil currently under threat of extinction. When researchers from the Conservancy and the Environmental Defense Fund analyzed this database they found that problems with alien invaders was the second most common reason for being on the list, affecting a staggering 49 percent of all species. Only human alteration of habitats turned out to be a bigger threat. There are no good studies outside the United States at present, but this ranking is probably a fair reflection of the problem right across the planet.

And the problem is not just a terrestrial one. Many aquatic organisms have been transplanted across the world's oceans by ships. Ocean-going vessels have been using water drawn into floodable holds and tanks for balance and stability since the 1880s. Often the water is taken on board at the place of departure and discharged at subsequent ports as cargo is loaded. James Carlton and Jonathan Geller studied the ballast tanks of ships arriving in Oregon from Japan and found 367 different types of animals and plants lurking inside. They concluded that entire coastal planktonic assemblages are being transported across oceans, and that bays, estuaries, and inland waters are consequently among the most threatened ecosystems in the world. The most notorious ballast-tank stowaway of recent years must be the Eurasian zebra mussel, which is currently causing havoc in North America. Zebra mussels reproduce rapidly, carpet lake and river beds, clog up municipal and industrial water inlets, disrupt algal populations, alter the concentration of nutrients in the water, and affect the functioning of whole ecosystems. The cost of clearing these animals from blocked intake pipes alone has been estimated at 2 billion U.S. dollars.

A particularly worrying aspect of alien invasions has just recently come to light. There is a widely held belief among botanists and ecologists that the impact of alien plants may be mitigated to some extent in species-rich native communities because diverse communities have a natural tendency to resist invasion. The most commonly offered justification for this view is that communities consisting of many different species tap important resources like water and nutrients in many different ways and thereby make more complete use of them. If essential resources are already being exploited to the full, then any species trying to invade from outside will, so to speak, find no room at the inn. If the invaders cannot establish themselves in the first place, then they cannot even threaten to oust the natives. The theory has always been controversial, with some evidence for and some against, but recent research in the United States has moved the debate in an unexpected and disquieting direction.

Thomas Stohlgren and his colleagues studied the relationship between the number of native plants and the number of exotic invaders on study sites across the Colorado Rockies and the grasslands of Wyoming, Colorado, South Dakota, and Minnesota. The results clearly show that communities rich in native plants in this part of the world are not resisting invasion by introduced species. In fact, the opposite seems to be true: more aliens have invaded the species-rich sites than those with fewer natives. Why this should be happening is unclear. Stohlgren notes that sites with many natives and aliens together tend to have the most fertile soils, suggesting that availability of nutrients and diversity might be linked in some way regardless of the provenance of the plants. Why this should be so is also far from clear, but there will undoubtedly be a flurry of research-grant applications in the near future from ecologists keen to find out.

Stohlgren states that his results are alarming, and for America's native flora this is certainly true, but the implications for our planet are potentially catastrophic. Unless the biosphere has some sort of natural immunity to our stirring, so that strongholds of native diversity will repel boarders, global mixing can be prevented only by affirmative action on our part. Even worse, if the natural tendency of alien species is to *congregate* in areas of high native diversity, as Stohlgren's work suggests, then homogenization of the biosphere could be proceeding many times faster than hitherto feared, and it may already be too late to do much about it.

The threat of global mixing is perhaps best appreciated by looking at current levels of biodiversity on different continents. Figure 9.1 shows the number of mammal species on Earth's five major continental landmasses. The typical relationship between area and species-richness is clearly evident, but the most interesting aspect of the figure is the dotted extension of the best-fit line which runs off to the right. This line stops at a point on the horizontal axis representing the area of all the continental landmasses combined. The value on the vertical axis suggests that if all the continents were joined together, the number of terrestrial mammal species on Earth would probably be around 2000, which is only half the number that actually exist today. Isolation breeds diversity by keeping creatures apart, and this rule holds as well for continents as it does for oceanic islands. Stirring the biosphere as we are currently doing is akin to pushing the continental landmasses together, and Figure 9.1 suggests that the effect on global biodiversity may be catastrophic.

The few scientists who study the problem of global mixing often find it frustratingly difficult to get others to appreciate the scale of the problem. The major stumbling block seems to be that most people have difficulty in imagining (or caring about) processes that operate over timescales greater than a century or so. This is hardly surprising, as our brains are not wired to appreciate millennia or eons any more than our eyes are wired to see atoms or distant galaxies. And many scientists seem to be just as temporally challenged as the rest of us. Colleagues in different disciplines tend to react predictably to the issue of global mixing: biologists express interest, ecologists express concern, and paleontologists intuitively grasp the gravity of the problem well before I have a chance to spell it out. The difference is one of training and perspective: paleontologists regularly deal with global changes over timescales longer than those relevant to the process of biospheric mixing, so they have an appropriate temporal framework

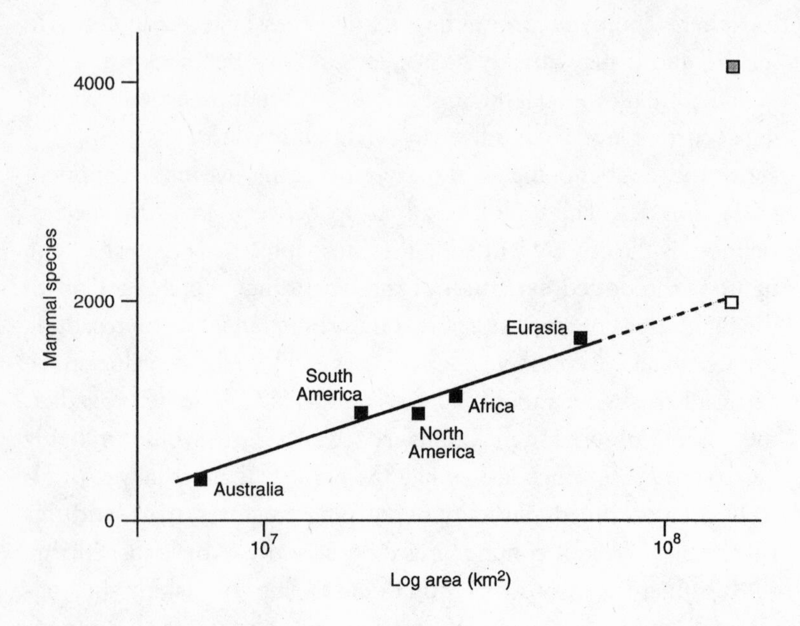

FIGURE 9.1 Adapted from Vitousek et al. (1997).

within which to evaluate the nature and extent of the threat. The four millennia during which humans have homogenized life on the world's oceanic islands, you can see them thinking, are just a geological microsecond, less than 0.0001 percent of the Cenozoic, and the Cenozoic is less than 0.02 percent of the history of life. From a geologic perspective, the rate at which we are breaking down the Earth's geographic barriers and homogenizing the biosphere is truly staggering.

In an attempt to express the potential seriousness of alien invasions, some scientists have drawn analogies with other environmental problems that have received much greater attention. The most common and useful analogy is with the greenhouse effect. As David Burney points out: "Biological invasion is really a bigger [problem] than a lot of the horrible things we hear about all the

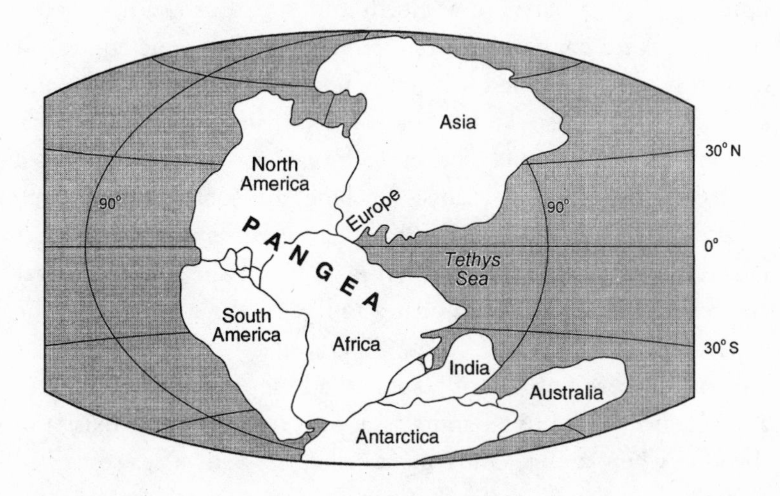

FIGURE 9.2 The position of the continents at the end of the Permian period.

time, like global warming. Many of the impacts of global warming could in time be reversed, but once you homogenise the biodiversity of the world, there's really no going back from that."[5] It is a reasonable comparison. Politicians are currently trying to limit the release of greenhouse gases into the atmosphere, so why should they ignore the equal or greater threat of biological invasions? But to those who study the deep history of life, Burney's words have an extra chilling resonance. History suggests, in fact, that the long-term impact of our actions on the biosphere may be more profound than anyone has hitherto realized. To understand why, we must delve into the fossil record one final time.

About 250 million years ago at the end of the Permian period (Fig. 3.1), the geography of the Earth was very different. All the major areas of continental crust had coalesced into a single landmass called Pangea (Fig. 9.2). On either side of the equator, cov-

5. Quoted in Holmes (1998).

ering much of what is now North America, Europe, and South America, were colossal, searing deserts. The Paleozoic equivalent of rain forests flourished in what is now China, the Malay peninsula, and Venezuela. To the south over Africa, India, Antarctica, and much of Australia, the climate was cooler and supported plains of cold-adapted plants. Girdling the southern pole were forests of the fernlike cold-weather plant *Glossopteris,* which must have dominated this part of Pangea in much the same way that conifers dominate northern climes today. At the other end of the supercontinent, in what is now Siberia and Mongolia, a similar cold-weather assemblage of plants called the Angara flora thrived. A single ocean, dubbed Panthalassa, surrounded Pangea except in the east, where a huge embayment called the Tethys Sea separated Eurasia from the southern continents.

Variation in climate maintained a variety of habitats arrayed between the equator and the poles, but with all ocean barriers eliminated, animals and plants had much easier access to terrestrial ecosystems around the planet. Richard Fortey (1997) sums up the likely consequences of the formation of Pangea for terrestrial life:

> The biological implications of having all the major land masses conjoined in the late Permian are profound. We are so used to our dispersed and scattered continents that we take the facts of isolation for granted. Imagine for a moment that all the continents were united together today. There would be a free-for-all as animals and plants dispersed in all directions; some would prosper, without doubt, but many others would not—consider, for example, how mankind's comparatively recent introduction of cats and foxes into Australia has had a devastating effect on dozens of the endemic marsupial species. . . . If everything could compete there would be more losers than winners. . . . In this respect, a world joined would be a world reduced.

Global mixing, in other words, 250 million years before human beings had the bright idea of lashing trees together into rafts. While Pangea existed, a great many families and even genera[6] of amphibians, reptiles, and early members of the lineage leading to mammals had worldwide distributions. Comparing the diversity of ancient faunas with modern ones is fraught with difficulty because ancient diversity has to be estimated from the remains of long-dead creatures encased in whatever rocks have managed to survive hundreds of millions of years of tectonism and weathering, but paleontologists are certain that tetrapods were much less diverse in the Permian than they are today, and that very similar types of creatures roamed the length and breadth of the Pangean supercontinent.

At the height of this continental amalgamation the biosphere suffered the greatest mass extinction in history. Many people are aware that the event that wiped out the dinosaurs was a planetwide disaster, but the Cretaceous debacle was a relatively minor affair compared with the global crisis at the end of the Permian. It has been estimated that 95 percent of all species on Earth died out. Even insects, traditionally the most resilient of land animals, rarely suffering extinctions even at the family level, lost seven whole orders. Crinoids (sea lilies) lost two subclasses. Trilobites disappeared completely, as did blastoid sea urchins and all the characteristic Paleozoic corals. Reefs did not reappear in the fossil record for another 7 to 8 million years, one of the longest periods without these distinctive marine communities in Earth's history. (Corals per se must have survived somewhere, but we haven't yet found them in rocks of this period.) Silica-

6. Similar species are grouped together to form a genus (plural genera). Genera are then grouped into families, families into orders, orders into classes, and classes into phyla. Because families contain more species than each of the genera of which they are composed, we would expect families, on average, to have wider geographic distributions than genera. Pan-continental families of animals are exceptionally rare, and pan-continental genera even more so.

shelled marine protozoa called radiolarians, whose dead bodies form a characteristic type of marine sediment called chert, suffered the only serious crisis in their long evolutionary history. Like reefs, cherts were not deposited again until the middle Triassic. Many brachiopods, the Paleozoic equivalents of bivalves (the common seashells found on beaches across the world), were either extinguished or reduced from groups containing a hundred or more species to just two or three. By the early Triassic the diverse shelly assemblages of the Permian seas had been replaced nearly everywhere by just a few species of bivalve.

On land the extinction was just as severe. In southern Pangea, the endless fernlike *Glossopteris* forests, which had stood for many millions of years, were suddenly replaced by a low-diversity assemblage of conifers and club mosses. Elsewhere, other types of peat-forming trees disappeared, leaving a coal gap in the fossil record lasting well into the middle Triassic. Just as in the oceans, the most extraordinary aspect of the Earth's flora at this time was its cosmopolitan nature. Vast expanses of the club moss *Pleuromeia* grew in virtually all coastal habitats, while in drier inland areas the conifer *Voltzia* and the seed fern *Dicroidea* grew at all latitudes. These pan-Pangean distributions both on land and in the sea suggest a very uniform climate across the whole planet, a hint about the cause of the disaster to which we will return shortly.

Tetrapods were particularly badly hit. Twenty-one terrestrial families passed away, a greater proportional loss than in the oceans. The extinction of land animals seems to have consisted of a prolonged decline in diversity over the last few million years of the Permian, culminating in a sharp event that wiped out many small omnivores, all large herbivores, gliding reptiles, and six of nine families of amphibians. Once again, the surviving communities were peculiar in being of very low diversity and exceptionally cosmopolitan. Over 90 percent of the earliest Triassic tetrapod fossil assemblages from the supercontinent of Pangea consist of

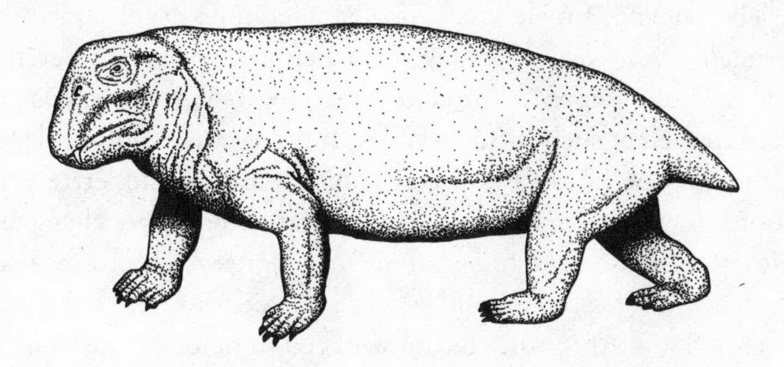

FIGURE 9.3 The early Triassic dicynodont therapsid *Lystrosaurus*. Length around 1 meter (39 inches).

just one genus of mammal-like reptile, the medium-sized dicynodont *Lystrosaurus* (Fig. 9.3).

What could have happened at the Permian-Triassic boundary to produce such wholesale devastation of life both on land and in the sea? What agent of destruction could have wiped out 95 percent of all species on our planet? The answer emerging from recent research is disquieting, given our current environmental predicament: it appears that the greatest disaster in the history of life was brought about by the combined effects of global mixing and global warming.[7]

---

7. Like most problems in paleontology, the cause(s) of the end-Permian extinction have been, and continue to be, the subject of intense debate. Numerous theories have been advanced over the years to explain the event, from cosmic radiation to falling sea levels. I agree with Hallam and Wignall (1997) that the evidence available at present points overwhelmingly toward global warming as the final and deadly coup de grâce for Paleozoic life. These authors also consider the old stalwart theories of sea-level drop (marine regression) and global refrigeration and dismiss both as either refuted or unsupported by the evidence. In brief, major marine regressions did occur in the Permian, but the youngest now appears to have ended several million years before the main extinction event. Regression may have been responsible for an earlier and lesser extinction episode, but the evidence now points strongly toward a sea-level rise across the Permian-Triassic boundary. The purported evidence for global refrigeration is the preferential loss of tropical species, the gradual nature of the extinction, the lack of

The evidence for elevated global temperatures at the end of the Permian is now overwhelming. Late Permian peats in Antarctica and Australia deposited around 70 degrees south are similar to those formed at such latitudes today, but they are rapidly replaced in the Triassic by soils indicating much warmer, temperate conditions. In Australia, the flora flipped from assemblages characteristic of present-day latitudes around 70 degrees south to those of much warmer regions around 40 to 58 degrees south. In the Karoo basin of South Africa the climate switched from temperate to semiarid. Add to these lines of evidence the fact that the earliest part of the Triassic is the only geologic interval in the last 600 million years for which there is no evidence anywhere of ice, and the case for planetwide warming seems irrefutable.

The warming of tropical regions has been estimated at around 6°C (11°F). The increase in temperature at higher latitudes would have been even greater, leading to a flattening of the temperature difference between the poles and the equator. The result was the complete loss of high-latitude floras and the establishment of a uniform warm-to-hot climate across most of the planet. The catastrophic decline in tetrapod diversity and the establishment of a single low-diversity assemblage of animals may have been partly due to increased environmental stress as arid conditions spread across the supercontinent, but the most important factor was probably the homogenization of habitats across Pangea. Not only were all the areas of continental crust on Earth joined together and thus easily navigable by walking animals, suddenly the pole-to-equator variation in habitat type began to fade

---

warm-water limestone formation in the early Triassic, the presence of deposits in Siberia indicating glacial activity and of deposits in Australia and Canada indicating cold conditions. However, the evidence now seems to support neither a gradual extinction nor preferential loss of tropical forms. And deposits of thick limestone do occur in many early Triassic low-latitude sites. The age of the Siberian rocks is also contested, and the Australian and Canadian rocks were deposited at high latitudes, so paleoevidence of cold temperatures is not surprising.

away too. Everywhere started to look the same, and the animals of the Earth naturally followed suit.[8]

Increasing temperatures and flattening of the pole-to-equator temperature gradient also had a drastic effect on the oceans. The solubility of oxygen in water falls as temperature rises, so global warming would have lowered the concentration of this essential element for life, but the most profound impact probably resulted, once again, from the equilibration of global temperatures. The pole-to-equator temperature gradient drives the circulation of the oceans. Flatten it, and ocean circulation slows down. Flatten it enough, and circulation could stop altogether. This is probably what happened at the end of the Permian. Unstirred, the oceans began to stagnate. Deep waters gradually lost oxygen, and species began to vanish.[9] Eventually the oceans stagnated to such an extent that nutrients in the upper part of the water column ran out. Planktonic plants cannot survive without nutrients, and without plants the whole marine ecosystem would have collapsed like a house of cards. Plankton productivity probably recovered quickly in the Triassic, but by then it was already too late to save the rich life of the Permian oceans.

What happened to cause such a severe episode of global warming? Worryingly, the same process that currently afflicts our planet: release of carbon dioxide into the atmosphere. Chemical

---

8. It has been suggested that most of the animals of this pandemic fauna, including the ubiquitous *Lystrosaurus,* were able to burrow and thereby hole up during the long, hot, dry seasons that must have gripped the planet.

9. The evidence for these oxygen-poor, or anoxic, conditions is abundant in the geologic and fossil record. Bottom-dwelling survivors of the Permian-Triassic boundary event tended to be species tolerant of low-oxygen conditions. Analysis of rock types across the boundary also indicates that the disappearance of the diverse late-Permian marine communities corresponded precisely with a change from intensely burrowed aerobic sea-floor sediments to finely laminated anaerobic rocks containing large amounts of pyrite, or fool's gold, a mineral indicative of anoxic conditions. Isotope analysis of pyrite in deep-sea boundary sediments from Japan suggests that they were deposited in anoxic waters similar to those of the modern-day Black Sea.

analyses of rocks suggests that CO_2 enrichment of the air began well before the final cataclysmic extinction. The most likely source of the extra carbon was the coal-bearing deposits of southern Pangea, which were uplifted by tectonic activity and oxidized, releasing large volumes of CO_2 into the atmosphere in the process. Whether this release was sufficient to elevate global temperatures by 6°C or more is debatable, but another massive dose of CO_2 was vented into the atmosphere over a period of around 900,000 years right at the Permian-Triassic boundary by the eruption in Siberia of 2 to 5 million cubic kilometers (0.5 to 1.2 million cubic miles) of basalt, the largest continental outpouring of volcanic material in the last 600 million years. This injection of volcanic CO_2 was probably the decisive event that ultimately tipped the biosphere into the new era of the Mesozoic.[10]

The disaster at the end of the Permian teaches many lessons about the processes we have set in train on our Cenozoic Earth. The late Permian had been a time of natural mixing as the continents finally came together to form the supercontinent of Pangea. This coalescence promoted extreme cosmopolitanism among animals and plants before the end of the period, but the establishment of pandemic, ultra-low-diversity communities both on land and in the sea during the Permian-Triassic transition was also directly related to the lack of geographic barriers. If the continents had been separated as they are today, it is inconceivable that the mammal-like reptile *Lystrosaurus,* the seed fern *Dicroidium,* and the bivalve *Claraia* would have

10. There is considerable debate about the effect(s) of the eruption of the Siberian Traps. Some believe, for example, that the eruptions were explosive enough to eject large volumes of dust into the atmosphere and cause a short period of global cooling, usually referred to as volcanic winter (e.g., Kozur, 1998). Others disagree (e.g., Hallam and Wignall, 1997; see Kozur for an extended treatment). Both parties do agree, however, that prolonged and intense global warming probably followed. Disagreement about the nature, timing, and extent of these environmental changes leads naturally to disagreement about the nature and timing of the kill mechanisms for different groups of marine and terrestrial animals and plants. Research continues.

come to dominate early Triassic terrestrial and oceanic ecosystems so completely. Diversity would have fallen precipitously, without doubt, but there would have been different *Lystrosaurus* analogues on different continents and many widely separated ocean basins and areas of continental shelf to protect and foster whatever marine diversity remained. Barriers isolate organisms from each other, and isolation is crucial for maintaining and promoting biodiversity, whatever the extraneous circumstances.

We may not be pushing the world's landmasses together in a physical sense, but our global wanderings are having much the same effect. In fact, the rate of biospheric mixing is greater now than at any other time in history: Pangea came together and then broke apart over a period of tens of millions of years, but we have breached the geographic barriers between continents *and* between continents and oceanic islands in just a few thousand. Everywhere is starting to look the same again, and this time it is happening in a geologic microsecond.

And the rate at which we are warming the atmosphere through the release of carbon dioxide and other greenhouse gases is just as alarming. Estimates of the increase in average global temperature over the next century range from 1 to 4.5°C (1.8 to 8.1°F) with most workers plumping for a figure somewhere in the middle.[11] Predictive models of climate change become increasingly unreliable as they are projected further into the future, but in almost all cases the models envisage rising temperatures for several hun-

---

11. Here I follow the official reports of the Intergovernmental Panel on Climate Change (IPCC). In the academic community the issue of global warming is much like that of evolution by natural selection: *everyone* seems to have an opinion on the subject. Many believe that the IPCC predictions should be taken with a pinch of salt. Others maintain that global warming isn't happening at all. Some believe that global warming could, in fact, trigger another glacial period. Yet others fear that the average temperature of the Earth could increase much more over the next few centuries than the IPCC worst-case scenario. Ironically, we probably understand more about the episode of global warming 250 million years ago than we understand about the present one.

dred years after 2100. Worst-case scenarios of increasing fossil-fuel consumption suggest a possible rise in average temperature of 5 to 10°C (9 to 18°F) by the middle of the twenty-second century.

We may be able to avoid such extreme increases if our lawmakers take steps to stabilize the concentration of greenhouse gases in the atmosphere. Let us hope that they do. Let us also hope that the world's volcanoes continue to behave themselves, and that the carbon currently locked up in woodlands and grasslands does not end up in the atmosphere when the former disappear and the latter turn into deserts, and that the predicted increase in the frequency and extent of wildfires does not have a similar effect, and that the rapid economic development of poor countries does not lead to the sort of profligate per capita combustion of fossil fuels characteristic of places like Britain and the United States, and so on through the maze of dimly perceived possibilities that our current climate models ignore.

A 3°C rise in average global temperature is only half that suggested for the Tethys region at the end of the Permian, but then we are only talking about a timescale of a hundred years. Even over this period, global warming could eliminate the Arctic ecosystem from Alaska and shift other plant associations 500 km (310 miles) to the north. The disappearance of the *Glossopteris* and Angaran floras from either end of Pangea at the close of the Permian provides obvious parallels. But the century over which scientists are predicting the impact of global warming does not even qualify as a tick of the geologic clock. Even the eruption of the Siberian Traps took 900,000 years. It probably took several million years for the Permian world to cook right through. Seen in the appropriate temporal context, a 3°C rise in average global temperature in a century is exceptional, to say the least.[12]

12. In the context of the last 6 million centuries this statement is surely true. However, the last 2 million years of Earth's history have been characterized by uncommonly

To contend that we are currently heading toward a Permian-like environmental crisis would make for a dramatic ending, but it would be an irresponsible conceit. The world is a very different place now. If all the continental landmasses were joined together and surrounded by a single ocean as they were in the Permian, the job of climate modelers would be a lot simpler, for a start. As it is, with our landmasses broken, widely scattered, and interspersed with ocean basins each with its own complicated patterns of circulation and interchange, predicting regional variations in the temperature and climate of our planet into the future has proved exceedingly difficult. And we also know that the Earth has experienced marked changes in climate at other times in the past—most notably during the ice ages of recent geologic history—without suffering anything like the wholesale decimation of life that characterized the Permian-Triassic transition. Perhaps the potential for global mixing at the end of the Permian was the key difference. Per-

---

variable climatic conditions, and there is some evidence that episodes of climate change in this period in some parts of the world may have been even more abrupt than those predicted for the twenty-first century (I chose the end-Permian as an analogue for our current situation rather than any other period of paleoclimatic change because this is the only other known example of simultaneous planetwide warming *and* mixing). Chemical analyses of ice cores from Greenland, for example, suggest that the temperature of this region fluctuated markedly during the last glacial period, with increases of as much as 7°C (13°F) in just a few decades. Some of the temperature changes in the preceding Eem interglacial suggested from analyses of these ice cores appear to be even more dramatic. It should be remembered, however, that these purported changes are for single points on the Earth's surface and should not be interpreted as variations in average global temperature. Moreover, some researchers believe that they are artifactual—that is, the product of deformation within the ice or the signatures of rapid changes in the chemical composition of snow. The latter could have resulted from changes in the temperature of source-water bodies or variation in the trajectory of airstreams feeding the Greenland ice cap.

Even assuming the Eem temperature fluctuations to be real, we can draw little comfort from them. The major factor distinguishing the Eem from today is that the former interglacial appears to have been slightly warmer. Human-caused global warming of just a few degrees, therefore, may trigger an episode of profound Eem-like climatic instability that could have dramatic effects on the biosphere. See Williams et al. (1998) for a critical review.

haps it was the sustained nature of the warming over several million years. Perhaps not. We simply do not understand the past or present behavior of our planet well enough to make trustworthy predictions about the consequences of our current experiments in planetary change. I have no idea whether the temperature of our Earth will stabilize at a noncatastrophic level (whatever that may be), nor of the extent to which global mixing and global warming may combine to accelerate the planetwide collapse in biodiversity that human population expansion has already precipitated, but as we continue to pump greenhouse gases into the atmosphere and stir up the biosphere, the only sure thing is that no one else knows either. So even if the end-Permian disaster is rejected as an appropriate analogue for our current environmental predicament, we should at least have the humility to take it as a warning.

"The present is the key to the past" runs a nineteenth-century geologic maxim. To this day teachers exhort their students to interpret the relics of the past in terms of processes operating in the modern world. For most purposes the maxim is a good one. Our hard-won knowledge of the Earth shows that it makes more sense, for instance, to interpret patterns of animal life in the Permian, Cretaceous, or Miocene in terms of ecological and evolutionary processes going on around us today, rather than as sequential whims of the Almighty. Indeed, most of the ideas and arguments put forward in this book, as in most works of historical science, are rooted firmly in this very philosophy. But a strong argument could also be made for taking seriously the idea that the past is the key to our future, and to that of all our relatives in the earthly family of life. Human memory and human history are just too fleeting to teach us much about the planet that is our shared inheritance. Only the rocks and bones are old enough to do this. Our species has matured to the point where we alone can read and understand these messages from the past. The time has come to pause awhile and take counsel.

KEY REFERENCES AND SELECTED FURTHER READING

I hope that the following list of key references and selected further reading will help general readers and students find additional information on the subjects covered in this book. The list comprises those texts and articles that I have found most useful; it is not intended to be comprehensive. References are listed in the order in which particular subjects and issues arise in the text. Wherever possible I have listed sources that should be accessible (both obtainable and understandable) to lay readers along with more technical books and articles. Where there was a choice of references giving the same or similar information, I have listed the most accessible.

CHAPTER ONE

Size, scaling, and energetics: Alexander (1992); McMahon and Bonner (1983); Schmidt-Nielsen (1984). **General animal physiology:** Louw (1993); Schmidt-Nielsen (1997); Withers (1992). **Elephant natural history:** Chadwick (1994); Schneck (1997); Shoshani and Tassy (1996). **Trunks as snorkels:** Gaeth et al. (1999). **Shivering pythons:** Secor and Diamond (1997). **Mammoths:** Lister and Bahn (1994).

CHAPTER TWO

Relationship between size and hairiness in savanna antelopes: Louw (1993). **Cardiac system of shrews:** Schmidt-Nielsen (1984). ***Batodonoides:*** Hecht (1998). **Lower size-limit for warm-blooded animals:** Schmidt-Nielsen (1984). **Hibernation:** Louw (1993); Calder (1994). **Energy-saving strategies:** McNab (1980); Montgomery, ed. (1978). **Naked mole-rats:** Lovegrove and Wissel (1988); Lovegrove (1989). **Insect physiology:** Heinrich (1993).

CHAPTER THREE

Differences in mammalian and reptilian metabolisms and the advantages of a high and constant body temperature: Withers (1992). **Arguments against**

the hot-is-good theory: Ruben (1995). **Body size reduction in the lineage leading to mammals:** McNab (1978). **Spotilla's mass-homeotherms:** Spotilla et al. (1973), (1991). **Aerobic-capacity theory for the evolution of warm-bloodedness:** Bennett and Ruben (1979); Bennett (1991); Ruben (1995). **Respiratory turbinates:** Hillenius (1992), (1994).

CHAPTER FOUR

Brontosaur renaissance: Bakker (1986). **Bakker's ideas:** Bakker (1972), (1975), (1980), (1986). **Making dinosaurs unextinct:** Bakker and Galton (1974); Bakker (1975), (1986). **Structure and function of mammalian and reptilian hearts:** Withers (1992). ***Sinosauropteryx, Caudiptedryx, and Protarchaeopteryx:*** Ackerman (1998); Chen et al. (1998); Qiang et al. (1998). **Bird-theropod link:** Dingus and Rowe (1998); Feduccia (1996); Padian and Chiappe (1998); Shipman (1998). **Polar dinosaurs:** Benton (1996); Clemens and Nelms (1993); Molnar and Wiffen (1994); Rich and Rich (1989). **Bone-texture studies and interpretations:** Bakker (1986); Farlow et al. (1995); de Ricqles (1980); Reid (1987), (1990); Ruben (1995). **Isotope analysis of dinosaur bones:** Barrick et al. (1996). **Bakker's predator–prey-ratio analyses:** Bakker (1975), (1986). **Arguments against warm-blooded dinosaurs:** Farlow et al. (1995); Farlow and Brett-Surman, eds. (1997); Ruben (1995); articles in Thomas and Olson, eds. (1980); and for a balanced overview, Fastovsky and Weishampel (1996). **CAT-scans of dino noses:** Ruben et al. (1996). **Piston ventilation system of *Sinosauropteryx:*** Ruben et al. (1997). **Piston ventilation system of *Scipionyx:*** Ruben et al. (1999). **Dinosaur extinction:** Alvarez (1997); Dingus and Rowe (1998); Fastovsky and Weishampel (1996). **Things dinosaurian:** Currie and Padian, eds. (1997); Farlow and Brett-Surman, eds. (1997).

CHAPTER FIVE

Latitudinal diversity gradients: Douglas (1998); Stiling (1999). **Effects of low temperature on cell physiology:** Fogg (1998). **Adaptations to cold:** Irving (1980); Schmidt-Nielsen (1997); Withers (1992). **Insulative quality of polar-bear fur:** Irving (1980); Louw (1993); Scholander et al. (1950). **White bears and black ravens:** Louw (1993). **Aquatic polar mammals:** Blix and Steen (1979); Irving (1980); Louw (1993); Schmidt-Nielsen (1997). **Leg fat in polar animals:** Irving (1980). **Thermoregulation in newborn polar animals:** Blix and Steen (1979); Whittow and Tazawa (1991). **Emperor penguins:** Blix and Steen

(1979); Fothergill (1993). **Spadefoot toads:** Flegg (1993). **Camels:** Flegg (1993); Schmidt-Nielsen et al. (1957), (1997). **Oryx:** Taylor (1980). **Deep-breathing oryx:** Louw (1993). **Jackrabbits:** Flegg (1993). **Northern Eurasian animals:** Sparks (1992). **Relationship between body size and diversity:** May (1978); McMahon and Bonner (1983).

CHAPTER SIX

Limits to the size of terrestrial reptiles: Paul (1998). **Benefits of warm-bloodedness at large body sizes:** Bakker (1980), (1986). **Ecological and geographical characteristics of extant mega-reptiles:** Bakker (1986). **Komodo dragons:** Auffenberg (1981); Green et al. (1991); Preston (1982); Quammen (1996). **Komodo pygmy elephants:** Diamond (1987). **Giant tortoises:** Alderton (1993); Bakker (1986). **Boid snakes:** Burton (1998); Seigel et al., eds. (1987); Seigel and Collins, eds. (1993); Weidensaul (1991). **Australia:** Flannery (1991), (1996); Milewski (1981), (1983); Archer et al. (1994); Wroe (1999). ***Megalania:*** Flannery (1996). **Pristichampsine and sebecosuchian crocodilians:** Langston (1975); Steel (1989). ***Quinkana:*** Flannery (1996); Salisbury and Willis (1996).

CHAPTER SEVEN

Biogeography of crocodilians and turtles: Alderton (1993); Behler and Behler (1998); Brochu (in press); Markwick (1998); Taplin and Grigg (1989). **Freshwater mammals:** Carwardine (1995b); Evans (1994); Neill (1971); Whitfield, ed. (1998). **Otters:** Chanin (1985); Harris (1968). **Snorkeling alligators:** Ross, ed. (1989). **Marine mammals and reptiles:** Waller, ed. (1996). **Extinct marine reptiles:** Callaway and Nicholls, eds. (1997). ***Suchomimus:*** Sereno et al. (1998). **Green and brown food chains and survival into the Cenozoic:** Levinton (1996); Sheehan and Hansen (1986). **Reptile-rich fauna of Australia:** Flannery (1991), (1996); Milewski (1981), (1983). **Primary productivities of different environments:** Stiling (1999); Whittaker (1975). **Comparison of terrestrial and oceanic productivities:** Colinvaux (1978), (1993). **Dissipation of energy along food chains and production efficiencies:** Krebs (1994). **Ecology of rivers:** articles in Calow and Petts, eds. (1992), (1994), (1996); Jeffries and Mills (1990). **Variability and unpredictability of rivers:** articles in Calow and Petts, eds. (1992), (1994), (1996); Land Resources Division (1976); Resh et al. (1988); Schumm and Winkley, eds. (1994). **Otter diet and movements:** Chanin (1985). **Crocodile movements and resistance to drought:** Steel (1989).

Crocodilian biogeography and global movement: Brochu (in press). **Finless-porpoise distribution and river dolphins:** Carwardine (1995b). **Persecution and decline of crocodilians:** Behler and Behler (1998); Neill (1971); Ross, ed. (1989).

CHAPTER EIGHT

Cenozoic proliferation of birds: Dingus and Rowe (1998); Feduccia (1996). **1000-km/hr ostrich:** Louw (1993). **Weight and flight:** Burton (1990); Ellington (1991); McMahon and Bonner (1983). **Size and wing loading:** McMahon and Bonner (1983). **Problems of maintaining level flight in still air:** McMahon and Bonner (1983). **Relative energetic cost of swimming, walking, and flying:** McMahon and Bonner (1983). **Success of mammals via their teeth:** Benton (1991). **Bulk processing of plant material by extant birds:** Witmer and Rose (1991). ***Pteranodon*'s *weight:*** McNeill Alexander (1989). **Flightlessness and giant ground birds of the Cenozoic:** Feduccia (1996). **Jaw mechanics of *Diatryma:*** Witmer and Rose (1991). **Modal weight for mammals and birds:** Maurer (1998). **Woodland birds in the Neotropical and Ethiopian regions and Quaternary refuges:** Brook and Birkhead, eds. (1991); Bush and Colinvaux (1990); Williams et al. (1998). **No bias toward Neotropical forests among mammals:** Eisenberg (1981). **Global patterns of bat diversity:** Findley (1993). **Formation of Pacific islands:** Dingus and Rowe (1998); Whittaker (1998). **Bird-species densities:** Carwardine (1995a). **Hawaiian honeycreepers:** Cox and Moore (1993); Gorman (1979); Raikow (1976). **Bird diversity on islands and recent extinctions:** Steadman (1995).

CHAPTER NINE

Timescale of human transoceanic rafting: Whittaker (1998). **Paleoavifauna of New Zealand and Madagascar:** Diamond (1992). **Global bird extinctions:** Whittaker (1998). **Impact of introduced mammals relative to other tetrapods:** Ebenhard (1988); Lever (1994). **Impact of introduced mammals:** Lever (1994). **St. Helena:** Wallace (1895). **Release of feral cats in Australia:** Twyford (1991). **Problem of global mixing:** Holmes (1998); Vitousek et al. (1997); Wilcove et al. (1998); Williamson (1999). **Mixing in the oceans:** Carlton and Geller (1993). **Aliens invade hot spots of native plant diversity:** Stohlgren et al. (1999). **Permian geography and biogeography:** Behrensmeyer et al. (1992); Fortey (1997); Hallam and Wignall (1997); Osborne and Benton (1996); Osborne and Tarling (1995). **End-Permian extinction characteristics and causes:**

Hallam and Wignall (1997); Kozur (1998); Labandeira and Sepkoski (1993); Osborne and Benton (1996); Osborne and Tarling (1995); Retallack (1999). **Human-caused global warming:** Houghton et al. (1990); Houghton et al., eds. (1996); Jepma and Munasinghe (1998); Williams et al. (1998) Harvey (1999). **5–10°C increase in average global temperature:** Houghton (1997).

BIBLIOGRAPHY

Texts written for a general readership carry the superscript[1]. Technical books and articles accessible with effort and/or some background reading carry the superscript[2]. Material written by and for professional scientists carries the superscript[3].

Ackerman, J. (1998). Dinosaurs take wing. *National Geographic*, **194**, 74–99.[1]

Alderton, D. (1993). *Turtles and Tortoises of the World.* Blandford Press: London.[1–2]

Alexander, R. McNeill (1989). *Dynamics of Dinosaurs and Other Extinct Giants.* Columbia University Press: New York.[1]

——— (1992). *Exploring Biomechanics: Animals in Motion.* Scientific American Library: New York.[2]

Alvarez, W. (1997). *T. Rex and the Crater of Doom.* Princeton University Press: New Jersey.[1]

Anderson, M. (1951). *A Geography of Living Things.* English Universities Press: London.[2–3]

Archer, M., Hand, S., and Godthelp, H. (1994). *Riversleigh: The Story of Animals in Ancient Rainforests of Inland Australia.* 2d ed. Reed Books: Sydney.[2]

Auffenberg, W. (1981). *The Behavioral Ecology of the Komodo Monitor.* University Presses of Florida: Gainesville.[2]

Baker, R. (1982). *Migration: Paths Through Space and Time.* Holmes and Meier: London.[2–3]

Bakker, R. (1972). Anatomical and ecological evidence of endothermy in dinosaurs. *Nature,* **238**, 81–5.[3]

——— (1975). Dinosaur renaissance. *Scientific American*, **232,** 58–78.[1–2]

——— (1980). Dinosaur heresy—dinosaur renaissance: why we need endothermic archosaurs for a comprehensive theory of bioenergetic evolution. In R. Thomas and E. Olson, eds., *A Cold Look at the Warm-Blooded Dinosaurs.* AAAS Selected Symposia Series 28. AAAS: Washington. 351–462.[3]

——— (1986). *The Dinosaur Heresies.* Penguin: London.[1]

Bakker, R., and Galton, P. (1974). Dinosaur monophyly and a new class of vertebrates, *Nature,* **248**, 168–72.[3]

Barnes, R., and Mann, K. (1980). *Fundamentals of Aquatic Ecosystems*. Blackwell Scientific Publications: Oxford.[2–3]

Barrick, R., Showers, W., and Fischer, A. (1996). Comparison of thermoregulation of four ornithischian dinosaurs and a varanid lizard from the Cretaceous Two Medicine formation: evidence from oxygen isotopes. *Palaios,* **11**, 295–305.[3]

Bauwens, D., Garland, T., Castilla, A., and van Damme, R. (1995). Evolution of sprint speed in the lacertid lizards: morphological, physiological and behavioral covariation. *Evolution,* **49**, 848–63.[3]

Behler, J., and Behler, D. (1998). *Alligators and Crocodiles*. Colin Baxter Photography: Grantown-on-Spey.[1]

Behrensmeyer, A., et al. (1992). *Terrestrial Ecosystems Through Time: Evolutionary Paleoecology of Terrestrial Plants and Animals*. University of Chicago Press: Chicago.[3]

Bellairs, A. (1969). *The Life of Reptiles*. Vols. 1 and 2. Weidenfeld and Nicholson: London.[2]

Bennett, A. (1991). The evolution of active capacity. *Journal of Experimental Biology,* **160**, 1–23.[3]

Bennett, A., and Ruben, J. (1979). Endothermy and activity in vertebrates. *Science,* **206**, 649–54.[3]

Benton, M. (1991). *The Rise of the Mammals*. Apple Press: London.[1]

——— (1996). *Historical Atlas of the Dinosaurs*. Penguin: London.[1]

——— (1997). *Vertebrate Palaeontology*. Chapman and Hall: London.[2–3]

Blix, A., and Steen, J. (1979). Temperature regulation in newborn polar homeotherms. *Physiological Reviews,* **59**, 285–304.[3]

Bradshaw, S. (1986). *Ecophysiology of Desert Reptiles*. Academic Press: Sydney.[3]

Bramwell, D., ed. (1979). *Plants and Islands*. Academic Press: London.[2–3]

Brenchley, P., ed. (1984). *Fossils and Climate*. Wiley: Chichester.[3]

Briggs. J. (1987). *Biogeography and Plate Tectonics*. Elsevier: Amsterdam.[2–3]

Brochu, C. (in press). Congruence between physiology, phylogenetics and the fossil record on crocodilian historical biography. In G. Grigg, ed., *Crocodilian Biology and Evolution*. Surrey Beatty and Sons: New South Wales.[3]

Brooke, M., and Birkhead, T. (1991). *The Cambridge Encyclopedia of Ornithology*. Cambridge University Press: Cambridge.[1]

Burton, J. (1998). *The Book of Snakes*. Eagle Editions: Hertfordshire.[1]

Burton, R. (1990). *Bird Flight*. Facts on File: New York.[2]

Busbey III, A. (1986). *Pristichampsus* cf. *P. vorax* (*Eusuchia; Prsitichampsinae*) from the Uintan of West Texas. *Journal of Vertebrate Paleontology,* **6**, 101–3.[3]

Bush, M., and Colinvaux, P. (1990). A pollen record of a complete glacial cycle from lowland Panama. *Journal of Vegetation Science,* **1**, 105–18.[3]

Calder, W. (1994). When do hummingbirds use torpor in nature? *Physiological Zoology,* **67**, 1051–76.[3]

Callaway, J., and Nicholls, E., eds. (1997). *Ancient Marine Reptiles.* Academic Press: San Diego.[3]

Calow, P., and Petts, G., eds. (1992). *The Rivers Handbook.* Vol. 1. Blackwell Scientific Publications: Oxford.[3]

——— (1994). *The Rivers Handbook.* Vol. 2. Blackwell Scientific Publications: Oxford.[3]

——— (1996). *River Biota.* Blackwell Scientific Publications: Oxford.[3]

Capula, M. (1990). *The Macdonald Encyclopedia of Amphibians and Reptiles.* Macdonald Orbis: London.[1]

Carlquist, S. (1974). *Island Biology.* Columbia University Press: New York.[2]

Carlton, J., and Geller, J. (1993). Ecological roulette: the global transport of nonindigenous marine organisms. *Science,* **261**, 78–82.[2-3]

Carroll, R. (1997). *Patterns and Processes of Vertebrate Evolution.* Cambridge University Press: Cambridge.[3]

Carwardine, M. (1995a). *The Guinness Book of Animal Records.* Guinness Publishing: Middlesex.[1]

——— (1995b). *Whales, Dolphins and Porpoises.* Dorling Kindersley: London.[1]

Censky, E., Hodge, K., and Dudley, J. (1998). Over-water dispersal of lizards due to hurricanes. *Nature,* **395**, 556.[1–2]

Chadwick, D. (1994). *The Fate of the Elephant.* Penguin: London.[1]

Chanin, P. (1985). *The Natural History of Otters.* Croom Helm: London.[2]

Chen, P., Dong, Z. and Zhen, S. (1998). An exceptionally well preserved theropod dinosaur from the Yixian formation of China. *Nature,* **391**, 147–57.[3]

Chinsamy, A. (1993). Image analysis and the physiological implications of the vascularization of femora in archosaurs. *Modern Geology,* **19**, 101–8.[3]

Coleman, N. (1991). *Encyclopedia of Marine Animals.* Blandford: London.[1]

Colinvaux, P. (1978). *Why Big Fierce Animals Are Rare: An Ecologist's Perspective.* Princeton University Press: New Jersey.[1]

——— (1993) *Ecology 2.* Wiley: New York.[2–3]

Collinson, A. (1988). *Introduction to World Vegetation.* Unwin Hyman: London.[2]

Coulson, R., Herbert, J., and Coulson, T. (1989). Biochemistry and physiology of alligator metabolism *in vivo. American Zoologist,* **29**, 921–34.[3]

Cox, C. (1974). Vertebrate palaeodistributional patterns and continental drift. *Journal of Biogeography,* **1**, 75–94.[3]

Cox, C., and Moore, P. (1993). *Biogeography: An Ecological and Evolutionary Approach*. Blackwell Scientific Publications: Oxford.[2]

Crawley, M. (1993). *Herbivory: The Dynamics of Animal-Plant Interactions.* Studies in Ecology 10. Blackwell Scientific Publications: Oxford.[3]

Currie, P., and Padian, K., eds. (1997). *Encyclopedia of Dinosaurs*. Academic Press: San Diego.[2]

Dal Sasso, C., and Signore, M. (1998). Exceptional soft-tissue preservation in a theropod dinosaur from Italy. *Nature,* **392**, 383–7.[3]

Darlington, P. (1957). *Zoogeography: The Geographical Distribution of Animals*. Wiley: New York.[2–3]

Diamond, J. (1974). Colonization of exploded volcanic islands by birds: the supertramp strategy. *Science,* **184**, 803–6[2-3]

——— (1987). Did Komodo dragons evolve to eat pygmy elephants? *Nature,* **326**, 832.[2]

——— (1992). *Rise and Fall of the Third Chimpanzee*. Vintage: London.[1]

Dingus, L., and Rowe, T. (1998). *The Mistaken Extinction: Dinosaur Evolution and the Origin of Birds*. W. H. Freeman: New York.[2]

Donnovan, S. (1989). *Mass Extinctions—Processes and Evidence*. Belhaven: London.[3]

Douglas, K. (1998). Hot spots. *New Scientist,* **158 (2128),** 32–6.[1]

Druett, J. (1983). *Exotic Intruders: The Introduction of Plants and Animals into New Zealand*. Heinemann: Auckland.[2]

Dutenhoffer, M., and Swanson, D. (1996). Relationship of basal to summit metabolic rate in passerine birds and the aerobic capacity model for the evolution of endothermy. *Physiological Zoology,* **69**, 1232–54.[3]

Ebenhard, T. (1988). Introduced birds and mammals and their ecological effects. *Swedish Wildlife Research,* **13, (4),** 5–107.[3]

Eisenberg, J. (1981). *The Mammalian Radiations: An Analysis of Trends in Evolution, Adaptation and Behaviour*. Athlone Press: London.[3]

Ellington, C. (1991). Limitations on animals' flight performance. *Journal of Experimental Biology,* **160**, 71–91.[3]

Elton, C. (1958). *The Ecology of Invasions by Plants and Animals*. Methuen: London.[2–3]

Ennis, C., and Marcus, N. (1996). *Biological Consequences of Climate Change.* University Science Books: California.[1–2]

Evans, P. (1994). *Dolphins*. Whittet Books: London.[1]

Farlow, J., Dodson, P., and Chinsamy, A. (1995). Dinosaur biology. *Annual Review of Ecology and Systematics,* **26**, 445–71.[3]

Farlow, J., and Brett-Surman, M., eds. (1997). *The Complete Dinosaur*. Indiana University Press: Bloomington.[2–3]

Fastovsky, D., and Welshampel, D. (1996).*The Evolution and Extinction of the Dinosaurs*. Cambridge University Press: Cambridge.[2]

Feduccia, A. (1996). *The Origin and Evolution of Birds*. Yale University Press: New Haven.[2]

Findley, J. (1993). *Bats: A Community Perspective*. Cambridge University Press: Cambridge.[3]

Flannery, T. (1991). The mystery of the Meganesian meat-eaters. *Australian Natural History,* **23**, 722–9.[1]

——— (1996). *The Future Eaters: An Ecological History of the Australian Lands and People*. Secker and Warburg: London.[1]

Flegg, J. (1993). *Deserts: Miracle of Life*. Blandford: London.[1]

Fogg, G. (1998). *The Biology of Polar Habitats*. Oxford University Press: Oxford.[2–3]

Forster, C., Sampson, S., Chiappe, L., and Krause, D. (1998). The theropod ancestry of birds: new evidence from the late Cretaceous of Madagascar. *Science,* **279**, 1915–19.[3]

Fortey, R. (1997). *Life, an Unauthorised Biography*. HarperCollins: London.[1]

Fothergill, A. (1993). *Life in the Freezer: A Natural History of the Antarctic*. BCA: London.[1]

Frith, H. (1979). *Wildlife Conservation*. Angus and Robertson: Sydney.[2–3]

Furley, P., and Newey, W. (1982). *Geography of the Biosphere: An Introduction to the Nature, Distribution and Evolution of the World's Life Zones*. Butterworth: London.[2]

Gaeth, A., Short, R., and Renfree, M. (1999). The developing renal, reproductive, and respiratory systems of the African elephant suggest an aquatic ancestry. *Proceedings of the National Academy of Sciences,* **96**, 5555–8.[3]

Gans, C. (1989). Crocodilians in perspective. *American Zoologist,* **29**, 1051–4.[2]

Garland, T., and Carter, P. (1994). Evolutionary physiology. *Annual Review of Physiology,* **56**, 579–621.[3]

Gleeson, T. (1991). Patterns of metabolic recovery from exercise in amphibians and reptiles. *Journal of Experimental Biology,* **160**, 187–207.[3]

Gorman, M. (1979). *Island Ecology*. Chapman and Hall: London.[2]

Gould, S. (1986). Play it again, life. *Natural History,* **2**, 18–26.[1]

Graham, A., and Beard, P. (1990). *Eyelids of Morning: The Mingled Destinies of Crocodiles and Men*. Chronicle Books: San Francisco.[1]

Green, B., King, D., Braysher, M., and Saim, A. (1991). Thermoregulation, water turnover and energetics of free-living Komodo dragons, *Varanus komodoensis*. *Comparative Biochemistry and Physiology,* **99A**, 97–101.[3]

Groves, R., and Di Castri, F. (1991). *Biogeography of Mediterranean Invasions*. Cambridge University Press: Cambridge.[2–3]

Hadley, N. (1972). Desert species and adaptation. *American Scientist,* **60**, 338–47.[3]

Hallam, A., and Wignall, P. (1997). *Mass Extinctions and Their Aftermath*. Oxford University Press: Oxford.[3]

Harris, C. (1968). *Otters*. Weidenfeld and Nicholson: London.[3]

Harvey, D. (1999). *Global Warming: The Hard Science*. Pearson Education: Harlow.[3]

Hayes, J., and Garland, T. (1995). The evolution of endothermy: testing the aerobic capacity model. *Evolution,* **49**, 836–47.[3]

Heaney, L., and Patterson, B. (1986). *Island Biogeography of Mammals*. Academic Press: London.[2-3]

Hecht, J. (1998). Small bite makes big impression. *New Scientist*, 10 October, 15.[1]

Heinrich, B. (1993). *The Hot-Blooded Insects: Strategies and Mechanisms of Thermoregulation*. Harvard University Press: Cambridge.[2]

Hengeveld, R. (1989). *The Dynamics of Biological Invasions*. Chapman and Hall: London.[3]

Hillenius, W. (1992). The evolution of nasal turbinates and mammalian endothermy. *Paleobiology,* **18**, 17–29.[3]

——— (1994). Turbinates in therapsids: evidence for late Permian origins of mammalian endothermy. *Evolution,* **48**, 207–29.[3]

Holmes, B. (1998). Day of the sparrow. *New Scientist*, 27 June, 32–5.[1]

Houghton, J. (1997). *Global Warming: The Complete Briefing*. Cambridge University Press: Cambridge.[2]

Houghton, J., Jenkins, G., and Ephraums, J., eds. (1990). *Climate Change: The Intergovernmental Panel on Climate Change Scientific Assessment*. Cambridge University Press: Cambridge.[2]

Houghton, J., Meira Filho, L., Callander, B., Harris, N., Kattenberg, A., and Maskell, K., eds. (1996). *Climate Change 1995: The Science of Climate Change*. Cambridge University Press: Cambridge.[2]

Irving, L. (1980). Adaptations to cold. In *Vertebrates: Physiology*. W. H. Freeman: San Francisco. pp. 155–60.[1–2]

Jefferies, M., and Mills, D. (1990). *Freshwater Ecology: Principles and Applications*. Belhaven Press: London.[2–3]

Jepma, C., and Munasinghe, M. (1998). *Climate Change Policy: Facts, Issues and Analyses*. Cambridge University Press: Cambridge.[3]

Jones, J. and Lindstedt, S. (1993). limits to maximal performance. Annual Review of Physiology, **55**, 547–69.[3]

Keast, A. (1972). *Mammals, Evolution and the Southern Continents*. State University of New York: Stony Brook.[2–3]

——— (1981). *Ecological Biogeography of Australia*. 3 vols. Junk: The Hague.[3]

Kellman, M., and Tackaberry, R. (1997). *Tropical Environments: The Functioning and Management of Tropical Ecosystems*. Routledge: London.[2]

Kemp, T. (1969). On the functional morphology of the gorgonopsid skull. *Philosophical Transactions of the Royal Society of London B,* **256**, 1–83.[3]

Koopowitz, H., and Kaye, H. (1990). *Plant Extinction: A Global Crisis*. 2d ed. Christopher Helm: London.[2]

Kozur, H. (1998). Some aspects of the Permian-Triassic boundary (PTB) and of the possible causes for the biotic crisis around this boundary. *Palaeogeography, Palaeoclimatology, Palaeoecology,* **143**, 227–72.[3]

Krebs, C. (1994). *Ecology: The Experimental Analysis of Distribution and Abundance*. Harper and Row: New York.[3]

Labandeira, C., and Sepkoski, J. (1993). Insect diversity in the fossil record. *Science,* **261**, 310–15.[3]

Lack, D. (1971). *Ecological Isolation in Birds*. Harvard University Press: Cambridge.[3]

Land Resources Division (1976). Land Resources Study 24. Land Resources Division: Surrey.[3]

Langston, W. (1975). Ziphodont crocodiles: *Pristichampsus vorax* (Trowell), new combination, from the Eocence of North America. *Fieldiana: Geology,* **33**, 291–314.[3]

Laurin, M. (1998). New data on the cranial anatomy of *Lycaenops (Synapsida, Gorgonopsidae)*, and reflections on the possible presence of streptostyly in gonopsians. *Journal of Vertebrate Paleontology,* **18**, 765–76.[3]

Lever, C. (1994). *Naturalized Animals: The Ecology of Successfully Introduced Species*. T&AD Poyser: London.[1-2]

Levinton, J. (1996). Trophic group and the end-Cretaceous extinction: did deposit feeders have it made in the shade? *Paleobiology,* **22**, 104–12.[3]

Lister, A., and Bahn, P. (1994). *Mammoths*. Macmillan: New York.[1]

Louw, G. (1993). *Physiological Animal Ecology*. Longman Scientific and Technical: Harlow.[2–3]

Louw, G., and Seely, M. (1982). *Ecology of Desert Organisms*, Longman: Essex.[3]

Lovegrove, B. (1989). The cost of burrowing by the social mole rats (*Bathyergidae) Cryptomys damarensis* and *Heterocephalus glaber*: the role of soil moisture. *Physiological Zoology,* **62**, 449–69.[3]

Lovegrove, B., and Wissel, C. (1988). Sociality in mole rats: Metabolic scaling and the role of risk sensitivity. *Oecologia*, **74**, 600–06[3]

Markwick, P. (1998). Crocodilian diversity in space and time: the role of climate in paleoecology and its implication for understanding K/T extinctions. *Paleobiology,* **24**, 470–97.[3]

Maurer, B. (1998). The evolution of body size in birds. II. The role of reproductive power. *Evolutionary Ecology,* **12**, 935–44.[3]

May, R. (1978). The dynamics and diversity of insect faunas. In L. Mound and N. Waloff, eds., *Diversity of Insect Faunas*. Blackwell Scientific Publications: Oxford, pp. 188–204.[3]

McGowan, C. (1997). *The Raptor and the Lamb: Predators and Prey in the Living World*. Allen Lane: London.[1]

McMahon, T., and Bonner, J. (1983). *On Size and Life*. Scientific American Books: New York.[2]

McNab, B. (1978). The evolution of endothermy in the phylogeny of mammals. *American Naturalist,* **112**, 1–21.[3]

——— (1980). Food habits, energetics, and the population biology of mammals. *American Naturalist,* **116**, 106–124.[3]

Mielke, H. (1989). *Patterns of Life: Biogeography of a Changing World*. Unwin Hyman: London.[2]

Milewski, A. (1981). A comparison of reptile communities in relation to soil fertility in the Mediterranean and adjacent arid parts of Australia and southern Africa. *Journal of Biogeography,* **8**, 493–503.[3]

——— (1983). A comparison of ecosystems in Mediterranean Australia and southern Africa: nutrient-poor sites at the Barrens and the Caledon Coast. *Annual Review of Ecology and Systematics.* **14**, 57–76.[3]

Montgomery, G., ed. (1978). *The Ecology of Arborial Folivores*, Smithsonian Institution Press: Washington.[3]

Neill, W. (1971). *The Last of the Ruling Reptiles: Alligators, Crocodiles and Their Kin*. Columbia University Press: New York.[2]

Newbigin, M. (1968). *Plant and Animal Geography*, Methuen: London.[2]

Nitecki, M., ed. (1984). *Extinctions*. Chicago University Press: Chicago.[3]

Norman, D. (1994). *Prehistoric Life: The Rise of the Vertebrates*. Macmillan: New York.[1]

Oppenheimer, M. (1996) Appendix: Vulnerable Ecosystems. In Ennis, C., and Marcus, N., *Biological Consequences of Climate Change.* University Science Books: Sausalito.[2]

Osborne, R., and Benton, M. (1996). *The Viking Atlas of Evolution.* Viking: London.[1]

Osborne, R., and Tarling, D. (1995). *The Viking Historical Atlas of the Earth.* Viking: London.[1]

Padian, K., and Chiappe, L. (1998). The origin of birds and their flight. *Scientific American,* **278**, 28–37.[1]

Paul, G. (1998). Terramegathermy and Cope's Rule in the land of titans. *Modern Geology,* **23**, 179–217.[3]

Piazzini, G. (1960). *The Children of Lilith.* Dutton: New York.[2]

Preston, D. (1982). Komodo dragon. *Natural History,* **91**, 72–75.[1]

Quammen, D. (1996). *The Song of the Dodo: Island Biogeography in an Age of Extinctions.* Hutchinson: London.[1]

Qiang, J., Currie, P., Norell, M., and Shu-An, J. (1998). Two feathered dinosaurs from northeastern China. *Nature,* **393**, 753–61.[3]

Racey, P., and Swift, S., eds. (1995). *Ecology, Evolution and Behavior of Bats.* Proceedings of a Symposium held by the Zoological Society of London and the Mammal Society. London, 1993. Clarendon Press: Oxford.[3]

Raikow, R. (1976). The origin and evolution of the Hawaiian honeycreepers *(Drepanididae). Living Bird,* **15**, 95–117.[3]

Raup, D. (1979). Size of the Permo-Triassic bottleneck and its evolutionary implications. *Science,* **206**, 217–8.[3]

Reid, R. (1987). Bone and dinosaur endothermy. *Modern Geology,* **11**, 133–54.[3]

——— (1990). Zonal 'growth rings' in dinosaurs. *Modern Geology,* **15**, 19–48.[3]

Resh, V., Brown, A., Covich, A., Gurtz, M., Li, H., Minshall, G., Reice, S., Sheldon, A., Wallace, J., and Wissmar, R. (1988). The role of disturbance in stream ecology. *Journal of the North American Benthological Society,* **7**, 433–55.[3]

Retallack, G. (1999). Postapocalyptic greenhouse palaeoclimate revealed by earliest Triassic paleosols in the Sydney basin, Australia. *Geological Society of America Bulletin,* **111**, 52–70.[3]

Ross, C., ed. (1989). *Crocodiles and Alligators.* Merehurst Press: London.[1]

Ruben, J. (1995). The evolution of endothermy in mammals and birds: from physiology to fossils. *Annual Review of Physiology,* **57**, 69–95.[3]

Ruben, J., Hillenius, W., Geist, N., Leitch, A., Jones, T., Currie, P., Horner,

J., and Espe, J. (1996). The metabolic status of some late Cretaceous dinosaurs. *Science,* **273**, 1204–7.[3]

Ruben, J., Jones, T., Geist, J., and Hillenius, W. (1997). Lung structure and ventilation in theropod dinosaurs and early birds. *Science,* **278**, 1267–70.[3]

Ruben, J., Dal Sasso, C., Geist, N., Hillenius, W., Jones, T., and Signore, M. (1999). Pulmonary function and metabolic physiology of theropod dinosaurs. *Science,* **283**, 514–16.[3]

Salisbury, S., and Willis, P. (1996). A new crocodilian from the early Eocene of southeastern Queensland and a preliminary investigation of the phylogenetic relationships of crocodyloids. *Alcheringa,* **20**, 179–226.[3]

Schmidt-Nielsen, K. (1984). *Scaling: Why Is Animal Size So Important?* Cambridge University Press: Cambridge.[2]

——— (1997). *Animal Physiology: Adaptation and Environment*. Cambridge University Press: Cambridge.[2–3]

Schmidt-Neilsen, K., Schmidt-Neilsen, B., Jarnum, S., and Houpt, J. (1957). Body temperature of the camel and its relation to water economy. *American Journal of Physiology,* **188**, 103–12.[3]

Schneck, M. (1997). *Elephants: Gentle Giants of Africa and Asia*. Parkgate Books: London.[1]

Scholander, P., Walters, V., Hock, R., Johnson, F., and Irving, L. (1950). Body insulation of some arctic and tropical mammals and birds. *Biological Bulletin,* **99**, 225–36.[3]

Schumm, S., and Winkley, B., eds. (1994). *The Variability of Large Alluvial Rivers*. ASCE Press: New York.[3]

Scientific American (1980). *Vertebrates: Physiology*. W. H. Freeman: San Francisco.[1–2]

Secor, S., and Diamond, J. (1997). Determinants of the post-feeding metabolic response of Burmese pythons, *Python molurus*. *Physiological Zoology,* **70**, 202–12.[3]

Seigel, R., and Collins J., eds. (1993). *Snakes: Ecology and Behavior*. McGraw-Hill: New York.[3]

Seigel, R., Collins, J., and Novak, S., eds. (1987). *Snakes: Ecology and Evolutionary Biology*. McGraw-Hill: New York.[3]

Sereno, P., Beck, A., Duthell, D., Gado, B., Larsson, H., Lyon, G., Marcot, J., Rauhut, O., Sadlier, R., Sidor, C., Varricchio, D., Wilson, G., and Wilson, J. (1998). A long-snouted predatory dinosaur from Africa and the evolution of spinosaurids. *Science,* **282**, 1298–1302.[3]

Sheehan, P., and Fastovsky, D. (1992). Major extinctions of land-dwelling vertebrates at the Cretaceous-Tertiary boundary, eastern Montana. *Geology,* **20**, 556–60.[3]

Sheehan, P., and Hansen, T. (1986). Detritus feeding as a buffer to extinction at the end of the Cretaceous. *Geology,* **14**, 868–70.[3]

Shine, R., and Madsen, T. (1996). Is thermoregulation unimportant for most reptiles? An example using water pythons (*Liasis fuscus*). *Physiological Zoology,* **69**, 252–69.[2–3]

Shipman, P. (1998). *Taking Wing: Archaeopteryx and the Evolution of Bird Flight.* Weidenfield and Nicholson: London.[1–2]

Shoshani, T., and Tassy, P. (1996). *The Proboscidea.* Oxford University Press: Oxford.[2–3]

Skelton, P., ed. (1993). *Evolution:A Biological and Palaeontological Approach.* Addison-Wesley: Wokingham.[2–3]

Smith, R. (1995). Changing fluvial environments across the Permian-Triassic boundary in the Karoo Basin, South Africa and possible causes of tetrapod extinctions. *Palaeogeography, Palaeoclimatology, Palaeoecology,* **117**, 81–104.[3]

Sparks, J. (1992). *Realms of the Russian Bear.* BBC Books: London.[1]

Spotilla, J., Lommen, P., Bakken, G., and Gates, D. (1973). A mathematical model for body temperatures of large reptiles: Implications for dinosaur ecology. *American Naturalist,* **107**, 391–404.[3]

Spotilla, J., O'Connor, M., Dodson, P., and Paladino, F. (1991). Hot and cold running dinosaurs: body size, metabolism and migration. *Modern Geology,* **16**, 203–27.[3]

Stanley, S. (1986). *Earth and Life Through Time.* W. H. Freeman: San Francisco.[2–3]

Steadman, D. (1995). Prehistoric extinctions of Pacific island birds: biodiversity meets zooarchaeology. *Science,* **267**, 1123–31.[3]

Steel, R. (1989). *Crocodiles.* Christopher Helm: London.[1–2]

Stiling, P. (1999). *Ecology: Theories and Applications.* 3rd ed. Prentice Hall: Saddle River.[3]

Stohlgren, T., Binkley, D., Chong, G., Kalkhan, M., Scheil, L., Bull, K., Otsuki, Y., Newman, G., Bashkin, M., and Son, Y. (1999). Exotic plant species invade hot spots of native plant diversity. *Ecological Monographs,* **69**, 24–46.[2–3]

Stonehouse, B. (1971). *Animals of the Arctic: The Ecology of the Far North.* Ward Lock: London.[2]

Sues, H.-D., and Boy, J. (1988). A procynosuchid cynodont from central Europe. *Nature,* **331**, 523–4.[3]

Taplin, L., and Grigg, G. (1989). Historical zoogeography of the eusuchian crocodilians: a physiological perspective. *American Zoologist,* **29**, 885–901.[3]

Tarling, D. (1992). *Plate Tectonics and Biological Evolution*. Carolina Biological Supply Company: Burlington.[2]

Taylor, C. (1980). The eland and the oryx. In *Verbetrates: Physiology*. W. H. Freeman: San Francisco. pp. 124–32.[1–2]

Thapar, V. (1997). *Land of the Tiger: A Natural History of the Indian Subcontinent*. BBC Books: London.[1]

Thomas, R., and Olson, E., eds. (1980). *A Cold Look at the Warm Blooded Dinosaurs*. AAAS Selected Symposia Series 28. AAAS: Washington.[2–3]

Tivy, J. (1998). *Biogeography. A Study of Plants in the Ecosphere*. Addison Wesley Longman: Essex.[2]

Townsend, C. (1980). *The Ecology of Streams and Rivers*. Studies in Biology No. 122. Edward Arnold Biology: London.[3]

Twyford, G. (1991). *Australia's Introduced Animals and Plants*. Reed Books: Balgowlah.[2]

Vitousek, P., D'Antonio, C., Loope, L., Reimanek, M., and Westbrooks, R. (1997). Introduced species: a significant component of human-caused global change. *New Zealand Journal of Ecology,* **21**, 1–16.[3]

Wallace, A. (1895). *Island Life*. Macmillan: London.[2]

Waller, G., ed. (1996). *Sealife: A Complete Guide to the Marine Environment*. Pica Press: Sussex.[1]

Weidensaul, S., (1991). *Snakes of the World*. Grange Books: London.[1]

Whitfield, P., ed. (1998). *The Illustrated Encyclopedia of Animals*. Marshall Publishing: London.[1]

Whittaker, R. (1975). *Communities and Ecosystems*. MacMillan: New York.[3]

——— (1998). *Island Biogeography: Ecology, Evolution and Conservation*. Oxford University Press: Oxford.[2–3]

Whittow, G., and Tazawa, H. (1991). The early development of thermoregulation in birds. *Physiological Zoology,* **64**, 1371–90.[3]

Wilcove, D., et al. (1998). Quantifying threats to imperiled species in the United States. *Bioscience,* **48**, 607–15.[2–3]

Williams, M., Dunkerly, D., de Decker, P., Kershaw, P., and Chappell, J. (1998). *Quaternary Environments*. Arnold: London.[2–3]

Williamson, M. (1986). *Quantitative Aspects of the Ecology of Biological Invasions*. Royal Society: London.[3]

——— (1999). Invasions. *Ecography,* **22**, 5–12.[3]

Willis, P., and Mackness, B. (1996). *Quinkana babarra*, a new species of ziphodont crocodile from the early Pliocene Bluff Downs Local Fauna, Northern Australia, with a revision of the genus. *Proceedings of the Linnean Society of New South Wales,* **116**, 143–51.[3]

Wilson, J. (1990). *Lemurs of the Lost World*. Impact Books: London.[1]

Withers, P. (1992). *Comparative Animal Physiology*. Saunders College Publishing: Fort Worth.[3]

Witmer, L., and Rose, K. (1991). Biomechanics of the jaw apparatus of the gigantic Eocene bird *Diatryma*: implications for diet and mode of life. *Paleobiology,* **17**, 95–120.[3]

Wroe, S. (1998). Killer kangaroo. *Australiasian Science,* **19**, 25–8.[2]

——— (1999). Killer kangaroos and other murderous marsupials. *Scientific American*, May, 58–64.[1]

INDEX

Page references in italics denote illustrations.